T0073974

SpringerBriefs in Applied Sciences and Technology

Series editor

Janusz Kacprzyk, Polish Academy of Sciences, Systems Research Institute, Warsaw, Poland

SpringerBriefs present concise summaries of cutting-edge research and practical applications across a wide spectrum of fields. Featuring compact volumes of 50 to 125 pages, the series covers a range of content from professional to academic. Typical publications can be:

- A timely report of state-of-the art methods
- An introduction to or a manual for the application of mathematical or computer techniques
- A bridge between new research results, as published in journal articles
- A snapshot of a hot or emerging topic
- An in-depth case study
- A presentation of core concepts that students must understand in order to make independent contributions

SpringerBriefs are characterized by fast, global electronic dissemination, standard publishing contracts, standardized manuscript preparation and formatting guidelines, and expedited production schedules.

On the one hand, **SpringerBriefs in Applied Sciences and Technology** are devoted to the publication of fundamentals and applications within the different classical engineering disciplines as well as in interdisciplinary fields that recently emerged between these areas. On the other hand, as the boundary separating fundamental research and applied technology is more and more dissolving, this series is particularly open to trans-disciplinary topics between fundamental science and engineering.

Indexed by EI-Compendex and Springerlink

More information about this series at http://www.springer.com/series/8884

Pratima Bajpai

Anaerobic Technology in Pulp and Paper Industry

 Springer

Pratima Bajpai
Pulp and Paper Consultant
Kanpur
India

ISSN 2191-530X   ISSN 2191-5318 (electronic)
SpringerBriefs in Applied Sciences and Technology
ISBN 978-981-10-4129-7   ISBN 978-981-10-4130-3 (eBook)
DOI 10.1007/978-981-10-4130-3

Library of Congress Control Number: 2017933673

Printed on acid-free paper

This Springer imprint is published by Springer Nature
The registered company is Springer Nature Singapore Pte Ltd.
The registered company address is: 152 Beach Road, #21-01/04 Gateway East, Singapore 189721, Singapore

Preface

The application of anaerobic technology in pulp and paper industry is gaining acceptance as a cost-effective treatment alternative. Compared to conventional aerobic methods, the anaerobic wastewater treatment concept offers a number of important benefits. These include lower energy requirements and operating costs as well as production of a useful energy by-product in the form of methane gas. Additionally, anaerobic treatment systems reduce considerably the volume of excess sludge produced due to the low cell yields of anaerobic bacteria. The low excess sludge production makes anaerobic treatment methods particularly attractive since waste sludge disposal is becoming a major problem for aerobic treatment systems. The low nutrient requirements of anaerobic bacteria is also an advantage in the treatment of nutrient deficient wastewaters such as those from pulp and paper mills. Furthermore, anaerobic treatment methods can potentially be combined with post-treatment methods by which valuable products like ammonia or sulphur can be recovered.

Anaerobic treatment of pulp and paper wastewater is now applied in several pilot and full-scale plants as alternative to aerobic treatment. Development of the various high-rate anaerobic processes and much more concentrated pulp mill effluents (due to extensive water recycling) make the economic benefit from anaerobic treatment more significant which in turn increases the interest in the use of this technology. This e-book presents the state-of-the-art report on treatment of pulp and paper industry effluents with anaerobic technology. Coverage ranges from basic reasons for anaerobic treatment, comparison between anaerobic and aerobic treatment, effluent types suitable for anaerobic treatment, design considerations for anaerobic treatment, anaerobic reactor configurations applied for treatment of pulp and paper industry effluents, present status of anaerobic treatment in pulp and paper industry, economic aspects, examples of full-scale installations and future trends.

Kanpur, India   Pratima Bajpai

Acknowledgements

Some excerpts taken from Bajpai Pratima (2000). Treatment of Pulp and Paper Mill Effluents with Anaerobic Technology, PIRA International, UK with kind permission from Smithers Pira, UK, the worldwide authority on the Packaging, Print and Paper supply chains.

Some excerpts taken from Bajpai Pratima (2013). Bleach plant effluents from pulp and paper industry. SpringerBriefs in Applied sciences and Technology, Springer International Publishing, DOI 10.1007/978-3-319-00545-4 with kind permission from Springer International Publishing.

Contents

1 General Background 1
References 4

2 Basics of Anaerobic Digestion Process 7
References 11

3 Process Parameters Affecting Anaerobic Digestion 13
3.1 Anaerobic Conditions 13
3.2 Temperature 14
3.3 pH and Alkalinity 15
3.4 Inhibitory Compounds 17
    3.4.1 Sulphur Compounds 18
    3.4.2 Hydrogen Peroxide 19
    3.4.3 Low-Molecular-Weight-Organic Compounds 19
    3.4.4 Heavy Metals 20
    3.4.5 Molecular Hydrogen 20
    3.4.6 Wood Constituents 20
    3.4.7 DTPA 21
3.5 Nutrients and Trace Elements 21
3.6 Solids Retention Time (SRT) 22
3.7 Volatile Solids Loading Rate 23
3.8 Hydraulic Retention Time 24
3.9 Mixing 24
References 25

4 Comparison of Aerobic Treatment with Anaerobic Treatment 29
References 33

5 Anaerobic Reactors Used for Waste Water Treatment 37
5.1 Anaerobic Lagoon or Covered Lagoon Reactor 38
5.2 Anaerobic Contact Reactor 39

5.3   Upflow Anaerobic Sludge Blanket Reactor   41
5.4   Anaerobic Filter Reactor.   43
5.5   Anaerobic Fluidized and Expanded Bed Systems   45
5.6   Anaerobic Baffled Reactor   47
5.7   Anaerobic Membrane Reactor.   49
5.8   Hybrid Upflow Anaerobic Sludge Blanket/Anaerobic Filter   50
References.   51

6   **Pulp and Paper Making Process**   55
References.   58

7   **Wastewater and Sludge from Pulp and Paper Production**
Processes   61
References.   67

8   **Anaerobic Treatment of Pulp and Paper Industry Effluents**   69
8.1   Present Status   69
8.2   Manufacturers of Commercial Reactors for Waste Water
Treatment and Commercial Installations   78
References.   81

9   **Economic Aspects**   87
References.   89

10   **Conclusion and Future Perspectives**.   91
References.   93

Index   95

List of Figures

Figure 2.1   Anaerobic pathway 8
Figure 3.1   Effect of temperature on anaerobic activity 14
Figure 3.2   Relative activity of methanogens to pH. Based
on Mata-Alvarez 2002; Seadi 2008 16
Figure 4.1   Generalized comparison between aerobic and anaerobic
wastewater treatment in terms of the fate of organic carbon
[expressed as chemical oxygen demand (COD)] and energy
production/consumption and nutrient requirements (expressed
as N-requirements) [adapted from van Lier et al. (2008)] 33
Figure 5.1   Anaerobic lagoons. Based on mebig.marmara.edu.tr/Enve424/
Chapter7.pdf 38
Figure 5.2   Anaerobic contact process. Based on Agbalakwe (2011);
mebig.marmara.edu.tr/Enve424/Chapter7.pdf 41
Figure 5.3   UASB reactor. Based on Agbalakwe (2011) 42
Figure 5.4   Anaerobic filter. Based on Agbalakwe (2011); mebig.
marmara.edu.tr/Enve424/Chapter7.pdf 44
Figure 5.5   Fluidized bed reactor. Based on Agbalakwe (2011); mebig.
marmara.edu.tr/Enve424/Chapter7.pdf 45
Figure 5.6   EGSB reactor. Based on Wilson (2014) 47
Figure 5.7   IC reactor. Based on Wilson (2014) 48
Figure 5.8   Anaerobic baffled reactor. Based on Agbalakwe (2011). 48
Figure 5.9   Anaerobic membrane bioreactor. Based on Agbalakwe
(2011); mebig.marmara.edu.tr/Enve424/Chapter7.pdf. 49
Figure 5.10 Hybrid reactor. Based on Agbalakwe (2011); mebig.marmara.
edu.tr/Enve424/Chapter7.pdf 50
Figure 6.1   Schematic of pulp and paper production process 56

List of Tables

Table 1.1   Anaerobic treatment. 2
Table 1.2   Important conditions for efficient anaerobic treatment 2
Table 2.1   Steps involved in anaerobic oxidation of complex wastes 8
Table 3.1   Advantages of thermophilic process 15
Table 3.2   Disadvantages of thermophilic process 15
Table 4.1   Advantages of anaerobic process. 30
Table 4.2   Limitations of anaerobic processes 31
Table 4.3   Comparison between aerobic and anaerobic
            treatment processes 32
Table 5.1   Advantages and disadvantages of anaerobic lagoons 40
Table 5.2   Advantages of UASB reactors. 43
Table 5.3   Challenges of UASB reactors 43
Table 6.1   Major sources of water effluents 57
Table 7.1   Untreated effluent loads from pulp and paper manufacture 62
Table 7.2   Pollutants from various sources of pulping
            and papermaking. 63
Table 7.3   Typical wastewater generation and pollution load
            from pulp and paper industry 63
Table 7.4   Typical pollution load per ton of production (kg/ton) 64
Table 7.5   Typical characteristics of wastewater (mg/l)
            at different processes. 64
Table 7.6   Characteristics of wastewater (mg/l) at various pulp
            and paper processes. 65
Table 7.7   Characteristics of wastewater (mg/l) at various pulp
            and paper processes. 65
Table 7.8   Effluents used for anaerobic treatment. 66
Table 8.1   Reduction of pollutants in anaerobic-aerobic treatment
            of bleaching effluent 72

Table 8.2   Reduction of pollutants with ultrafiltration plus anaerobic/
             aerobic system and the aerated lagoon technique.   73
Table 8.3   Anaerobic treatment of pulp and paper mill wastewater   78
Table 8.4   Manufacturers of anaerobic digesters.   79
Table 8.5   Few examples of Using anaerobic technologies in the Pulp
             and Paper Industry.   80
Table 9.1   Cost benefit analysis of aerobic and anaerobic-aerobic
             treatment[a].   88
Table 9.2   Savings in operating cost per year at the Hylte Bruk
             and SAICA ANAMET plants compared to aerobic treatment
             of the same wastewater   89

Chapter 1
General Background

Abstract General background and introduction on anaerobic treatment technology is presented in this chapter.

Keywords Anaerobic digestion · Biogas · Fossil fuel · Methane · Hydrogen · Biological treatment · Heat · Waste streams · Power

Anaerobic digestion is a well-established biological process for converting carbon-rich feedstocks into biogas, which can be used to replace fossil fuels in heat and power generation and as a transportation fuel, facilitating the development of a sustainable energy supply (Hubbe et al. 2016; Zhang et al. 2015; Bialek et al. 2014; Li et al. 2014; Weiland 2010; Ziganshin et al. 2013; Al Seadi 2001). This process does not require any air or oxygen; converts biomass in waste streams into a renewable energy source, and it also contributes to the treatment of these waste streams (Table 1.1).

Anaerobic treatment technology has been receiving growing interest since its first application (Van Lier 2008). There is no need to pay for the pumping of air into the system in the anaerobic system and the amounts of sludge produced are usually less than in conventional aerated biological treatment systems (Maat and Habets 1987; Ashrafi et al. 2015; Kamali and Khodaparast 2015). Anaerobic processes generate gases such as methane, requiring their collection and safe disposal. Nevertheless, the retrieving and reuse of biogases such as methane and hydrogen as a source of energy during full-scale treatment operations can provide substantial economic benefits to the treatment plants. The methane and hydrogen either can be sold or they can be burnt for the generation of heat (Tabatabaei et al. 2010). The essential conditions for efficient anaerobic treatment are shown in Table 1.2.

Several reviews have been published on the treatment of pulp and paper mill effluents using anaerobic technology (Graves and Joyce 1994; Rintala and Puhakka 1994; Rajeshwari et al. 2000; Savant et al. 2006; Meyer and Edwards 2014; Kamali and Khodaparast 2015; Ali and Sreekrishnan 2001; Kosaric and BlaszczyK 1992). Several authors have reported evaluations of factors affecting the anaerobic treatment of pulp and paper mill wastewaters (Kortekaas et al. 1998; Bengtsson et al.

© The Author(s) 2017

P. Bajpai, *Anaerobic Technology in Pulp and Paper Industry*, SpringerBriefs in Applied Sciences and Technology, DOI 10.1007/978-981-10-4130-3_1

Table 1.1 Anaerobic treatment

Energy consumption
Low
Biogas production 0.05–0.10 kWh/kg COD
Sludge production
Low
0.0300.05 kg/kg COD, Market value
Foot print
Small
Compact designs available
Wilson (2014). www.seai.ie/…Energy…/Waste-to-Energy—Anaerobic-digestion-for-large-industry.p

Table 1.2 Important conditions for efficient anaerobic treatment

Absence of toxic/inhibitory compounds in the influent
Maintain pH in the neutral range - 6.8–7.2
Sufficient presence of alkalinity
Low volatile fatty acids
Temperature in the mesophilic range (30–38 °C)
Enough nutrients (nitrogen and phosphorous) and trace metals especially, Fe, Co, Ni, etc. COD:N:P: 350:7:1 (for highly loaded system)   1000:7:1 (lightly loaded system)
Avoidance of excessive air/oxygen exposure
Based on www.sswm.info/…/MANG%20ny%20Introduction%20in%20the%20technical%20des

2008; Sierraalvarez et al. 1991; Korczak et al. 1991; Vidal et al. 1997; Ruas et al. 2012; Krishna et al. 2014; Larsson et al. 2015).

Anaerobic digestion offers a platform for waste water treatment in terms of environmental management in addition to biogas production. The integrated biorefinery involving the conversion of biomass into biofuels, bio-based chemicals, biomaterials can also be developed and implemented based on anaerobic digestion (Uellendahl and Ahring 2010; Uggetti et al. 2014). Generally, in the context of the integration of forest biorefinery with traditional pulp and paper manufacturing processes, anaerobic digestion of organic wastes from these processes for biogas production would fit well into the biorefinery concept (Van Heiningen 2006; Amidon and Liu 2009; Jahan et al. 2013; Wen et al. 2013; Ahsan et al. 2014; Dansereau et al. 2014; Dashtban et al. 2014; Hou et al. 2014; Rafione et al. 2014; Wang et al. 2014; Liu et al. 2015; Matin et al. 2015; Oveissi and Fatehi 2014). The waste streams from the traditional pulp and paper making processes can be converted to valuable products by using anaerobic digestion. The use of anaerobic digestion process in the pulp and paper industry appears to be promising. In the other sectors also, the use of anaerobic digestion would create new possibilities.

By using anaerobic treatment instead of activated sludge about 1 kWh (fossil energy) kg-1 COD removed is saved, depending on the system which is used for aeration of activated sludge. Furthermore, under anaerobic conditions, the organic matter is converted in the gaseous energy carrier methane, producing about 13.5 MJ methane energy kg-1 COD removed, giving 1.5 kWh electric (assuming 40% electric conversion efficiency). In Netherlands, over 90% reduction in sludge production significantly contributed to the economics of the plant, whereas the high loading capacities of anaerobic high-rate reactors allowed for 90% reduction in space requirement, both compared to conventional activated sludge systems. These advantages resulted in the rapid development of anaerobic high-rate technology for industrial wastewater treatment. In this development, Dr. Lettinga group at Wageningen University, in close cooperation with the Paques BV and Biothane Systems International played a very important role (Lettinga 2014). Anaerobic high-rate technology has improved significantly in the last few decades with the applications of differently configured high-rate reactors, particularly for the treatment of industrial wastewaters. The rapid implementation of high-rate anaerobic treatment actually coincided with the implementation of the new environmental laws in Western Europe and the co-occurrence of very high energy prices in the 1970 s. High amounts of high strength wastewaters from distilleries, food processing and beverages industries, pharmaceutical industries, and pulp and paper industry required treatment. The first anaerobic full scale installations showed that during treatment of the effluents, significant amounts of useful energy in the form of biogas could be obtained for possible use in the production process (Van Lier 2008; Ersahin et al. 2007). The extremely low sludge production, was another very important advantage of high-rate anaerobic treatment systems. Interestingly, the production of granular sludge, gave a market value to excess sludge, as granular sludge is sold in the market for starting up new reactor. From the 1970s onwards, high-rate anaerobic treatment is particularly applied to organically polluted industrial wastewaters, which come from the agro-food sector and the beverage industries. Currently, in more than 90% of these applications, anaerobic sludge bed technology is used, for which the presence of granular sludge is of great importance. The number of anaerobic reactors installed and the application potential of anaerobic wastewater treatment is expanding rapidly. Currently, the number of installed anaerobic high rate reactors exceed 4000 (Van Lier et al. 2015). Nowadays wastewaters are treated that were earlier not considered for anaerobic treatment, such as wastewaters with a complex composition or chemical wastewaters containing toxic compounds. For the more extreme type of wastewaters novel high rate reactor system have been developed. Intensive pilot and laboratory studies and full-scale applications have demonstrated the suitability of anaerobic processes for the treatment of several types of pulp and paper industry wastewaters (Maat and Habets 1987; Korczak et al. 1991; Minami et al. 1991; Sierraalvarez et al. 1991; Vidal et al. 1997; Kortekaas et al. 1998; Ahn and Forster 2002; Buzzini and Pires 2002, 2007; Yilmaz et al. 2008; Tabatabaei et al. 2010; Lin et al. 2011; Saha et al. 2011; Elliott and Mahmood 2012; Bayr et al. 2013; Ekstrand et al. 2013; Hagelqvist 2013; Hassan et al. 2014; Meyer and Edwards 2014; Larsson et al. 2015).

References

Ahn JH, Forster CF (2002) A comparison of mesophilic and thermophilic anaerobic upflow filters treating paper-pulp-liquors. Process Biochem 38(2):257–262 (Article Number: PII S0032-9592 (02)00088-2)

Ahsan L, Jahan MS, Ni Y (2014) Recovering/concentrating of hemicellulosic sugars and acetic acid by nanofiltration and reverse osmosis from prehydrolysis liquor of kraft based hardwood dissolving pulp process. Bioresour Technol 155:111–115. doi:10.1016/j.biortech.2013.12.096

Al Seadi T (2001) Good practice in quality management of AD residues from biogas production, Report made for the International Energy Agency, Task 24-energy from biological conversion of organic waste, published by IEA Bioenergy and AEA Technology Environment, Oxfordshire, UK

Ali M, Sreekrishnan TR (2001) Aquatic toxicity from pulp and paper mill effluents: a review. Adv Environ Res 5(2):175–196. doi:10.1016/S1093-0191(00)00055-1

Amidon TE, Liu S (2009) Water-based woody biorefinery. Biotechnol Adv 27(5):542–550. doi:10.1016/j.biotechadv.2009.04.012

Ashrafi O, Yerushalmi L, Haghighat F (2015) Wastewater treatment in the pulp-and-paper industry: a review of treatment processes and the associated greenhouse gas emission. J Environ Manage 158:146–157. doi:10.1016/j.jenvman.2015.05.010

Bayr S, Kaparaju P, Rintala J (2013) Screening pretreatment methods to enhance thermophilic anaerobic digestion of pulp and paper mill wastewater treatment secondary sludge. Chem Eng J 223:479–486. doi:10.1016/j.cej.2013.02.119

Bengtsson S, Hallquist J, Werker A, Welander T (2008) Acidogenic fermentation of industrial wastewaters: effects of chemostat retention time and pH on volatile fatty acids production. Biochem Eng J 40(3):492–499. doi:10.1016/j.bej.2008.02.004

Bialek K, Cysneiros D, O'Flaherty V (2014) Hydrolysis, acidification and methanogenesis during low-temperature anaerobic digestion of dilute dairy wastewater in an inverted fluidised bioreactor. Appl Microbiol Biotechnol 98(2):8737–8750. doi:10.1007/s00253-014-5864-7

Buzzini AP, Pires EC (2002) Cellulose pulp mill effluent treatment in an upflow anaerobic sludge blanket reactor. Proc Biochem 38(5):707–713. doi:10.1016/S0032-9592(02)00190-5

Buzzini AP, Pires EC (2007) Evaluation of an upflow anaerobic sludge blanket reactor with partial recirculation of effluent used to treat wastewaters from pulp and paper plants. Bioresour Technol 98(9):1838–1848. doi:10.1016/j.biortech.2006.06.030

Dansereau LP, El-Halwagi M, Mansoornejad B, Stuart P (2014) Framework for margins-based planning: Forest biorefinery case study. Comput Chem Eng 63:34–50. doi:10.1016/j.compchemeng.2013.12.006

Dashtban M, Gilbert A, Fatehi P (2014) A combined adsorption and flocculation process for producing lignocellulosic complexes from spent liquors of neutral sulfite semichemical pulping process. Bioresour Technol 159:373–379. doi:10.1016/j.biortech.2014.03.006

Ekstrand E, Larsson M, Truong X, Cardell L, Borgström Y, Björn A, Ejlertsson J, Svensson BH, Nilsson F, Karlsson A (2013) Methane potentials of the Swedish pulp and paper industry—a screening of wastewater effluents. Appl Energy 112:507–517

Elliott A, Mahmood T (2012) Comparison of mechanical pretreatment methods for the enhancement of anaerobic digestion of pulp and paper waste activated sludge. Water Environ Res 84(6):497–505. doi:10.2175/106143012X13347678384602

Ersahin ME, Dereli RK, Insel G, Ozturk I, Kinaci C (2007) Model based evaluation for the anaerobic treatment of corn processing wastewaters. Clean-Soil Air Water 35(6):576–581

Graves JW, Joyce TW (1994) Critical review of the ability of biological treatment systems to remove chlorinated organics discharged by the paper industry. Water SA 20(2):155–160

Hagelqvist A (2013) Batchwise mesophilic anaerobic co-digestion of secondary sludge from pulp and paper industry and municipal sewage sludge. Waste Manage 33(4):820–824. doi:10.1016/j.wasman.2012.11.002

Hassan SR, Zwain HM, Zaman NQ, Dahlan I (2014) Recycled paper mill effluent treatment in a modified anaerobic baffled reactor: start-up and steady-state performance. Environ Technol 35 (3):294–299. doi:10.1080/09593330.2013.827222

Hou Q, Wang Y, Liu W, Liu L, Xu N, Li Y (2014) An application study of autohydrolysis pretreatment prior to poplar chemi-thermomechanical pulping. Bioresour Technol 169:155–161. doi:10.1016/j.biortech.2014.06.091

Hubbe MA, Metts JR, Hermosilla D, Blanco MA, Yerushalmi L, Haghighat F, Lindholm-Lehto P, Khodaparast Z, Kamali M, Elliott A (2016) Wastewater treatment and reclamation: a review of pulp and paper industry practices and opportunities. BioResources 11(3):7953–8091

Jahan MS, Rukhsana B, Baktash MM, Ahsan L, Fatehi P, Ni Y (2013) Pulping of non-wood and its related biorefinery potential in Bangladesh: a review. Curr Org Chem 17(15):1570–1576. doi:10.2174/13852728113179990065

Kamali M, Khodaparast Z (2015) Review on recent developments on pulp and paper mill wastewater treatment. Ecotoxicol Environ Safety 114:326–342. doi:10.1016/j.ecoenv.2014.05.005

Korczak MK, Koziarski S, Komorowska B (1991) Anaerobic treatment of pulp-mill effluents. Water Sci Technol 24(7):203–206

Kortekaas S, Sijngaarde RR, Clompt JW, Lettinga G, Field JA (1998) Anaerobic treatment of hemp thermomechanical pulping wastewater. Water Res 32(11):3362–3370. doi:10.1016/S0043-1354(98)00120-1

Kosaric N, BlaszczyK R (1992) Industrial effluent processing. encyclopedia of microbiology, vol 2 (Lederberg J. ed.). Academic Press Inc., New York, pp 473–491

Krishna KV, Sarkar O, Mohan SV (2014) Bioelectrochemical treatment of paper and pulp wastewater in comparison with anaerobic process: integrating chemical coagulation with simultaneous power production. Bioresour Technol 174:142–151. doi:10.1016/j.biortech.2014.09.141

Larsson M, Truong XB, Bjorn A, Ejlertsson J, Bastviken D, Svensson BH, Karlsson A (2015) Anaerobic digestion of alkaline bleaching wastewater from a kraft pulp and paper mill using UASB technique. Environ Technol 36(12):1489–1498. doi:10.1080/09593330.2014.994042

Lettinga G (2014) My anaerobic sustainability story. LeAF Publisher, Wageningen, p 200. http://www.leafwageningen.nl/en/leaf.htm

Li YF, Nelson MC, Chen PH, Graf J, Li Y, Yu Z (2014) Comparison of the microbial communities in solid-state anaerobic digestion (SS-AD) reactors operated at mesophilic and thermophilic temperatures. Appl Microbiol Biotechnol 99(2):969–980. doi:10.1007/s00253-014-6036-5

Lin Y, Wang D, Li Q, Xiao M (2011) Mesophilic batch anaerobic co-digestion of pulp and paper sludge and monosodium glutamate waste liquor for methane production in a bench-scale digester. Bioresour Technol 102(4):3673–3678. doi:10.1016/j.biortech.2010.10.114

Liu J, Li M, Luo X, Chen L, Huang L (2015) Effect of hot-water extraction (HWE) severity on bleached pulp based biorefinery performance of eucalyptus during the HWE-Kraft-ECF bleaching process. Bioresour Technol 18:183–190. doi:10.1016/j.biortech.2015.01.055

Maat DZ, Habets LHA (1987) Upflow anaerobic sludge blanket wastewater treatment system: technological review. Pulp Paper Canada 88(11):60–64

Matin M, Rahaman MM, Nayeem J, Sarkar M, Jahan MS (2015) Dissolving pulp from jute stick. Carbohydr Polym 115:44–48. doi:10.1016/j.carbpol.2014.08.090

Meyer T, Edwards EA (2014) Anaerobic digestion of pulp and paper mill wastewater and sludge. Water Res 65:321–349. doi:10.1016/j.watres.2014.07.022

Minami K, Okamura K, Ogawa S, Naritomi T (1991) Continuous anaerobic treatment of waste-water from a kraft pulp mill. J Ferment Bioeng 71(4):270–274. doi:10.1016/0922338X(91)90280-T

Oveissi F, Fatehi P (2014) Production of modified bentonite via adsorbing lignocelluloses from spent liquor of NSSC process. Bioresour Technol 174:152–158. doi:10.1016/j.biortech.2014.10.014

Rafione T, Marinova M, Montastruc L, Paris J (2014) The green integrated forest biorefinery: an innovative concept for the pulp and paper mills. Appl Therm Eng 73(1):72–79. doi:10.1016/j.applthermaleng.2014.07.039

Rajeshwari KV, Balakrishnan M, Kansal A, Lata K, Kishore VVN (2000) State of the art of anaerobic digestion technology for industrial wastewater treatment. Renew Sustain Energy Rev 4(2):135–156. doi:10.1016/S1364-0321(99)00014-3

Rintala JA, Puhakka JA (1994) Anaerobic treatment in pulp and paper mill waste management: review. Bioresour Technol 47(1):1–18. doi:10.1016/0960-8524(94)90022-1

Ruas DB, Chaparro TR, Pires EC (2012) Advanced oxidation process H_2O_2/UV combined with anaerobic digestion to remove chlorinated organics from bleached kraft pulp mill wastewater. Revista Facultad Ingen – Univ 63:43–54

Saha M, Eskicioglu C, Marin J (2011) Microwave, ultrasonic and chemomechanical pretreatments for enhancing methane potential of pulp mill wastewater treatment sludge. Bioresour Technol 102(17):7815–7826. doi:10.1016/j.biortech.2011.06.053

Savant DV, Abdul-Rahman R, Ranade DR (2006) Anaerobic degradation of adsorbable organic halides (AOX) from pulp and paper industry wastewater. Bioresour Technol 97(9):1092–1104. doi:10.1016/j.biortech.2004.12.013

Sierraalvarez R, Kortekaas S, Vaneckert M, Lettinga G (1991) The anaerobic biodegradability and methanogenic toxicity of pulping wastewaters. Water Sci Technol 24(3–4):113–125

Tabatabaei M, Rahim RA, Abdullah N, Wright ADG, Shirai Y, Sakai K, Sulaiman A, Hassan MA (2010) Importance of the methanogenic archaea populations in anaerobic wastewater treatments. Process Biochem 45(8):1214–1225. doi:10.1016/j.procbio.2010.05.017

Uellendahl H, Ahring BK (2010) Anaerobic digestion as final step of a cellulosic ethanol biorefinery: biogas production from fermentation effluent in a UASB reactor—pilot-scale results. Biotechnol Bioeng 107(1):59–64. doi:10.1002/bit.22777

Uggetti E, Sialve B, Trably E, Steyer JP (2014) Integrating microalgae production with anaerobic digestion: a biorefinery approach. Biofuel Bioprod Bior 8(4):516–529. doi:10.1002/bbb.1469

Van Heiningen ARP (2006) Converting a kraft pulp mill into an integrated forest biorefinery. Pulp Pap-Canada 107(6):38–43

Van Lier JB (2008) High-rate anaerobic wastewater treatment: diversifying from end-of-the-pipe treatment to resource oriented techniques. Water Sci Technol 57(8):1137–1148

Van Lier J, Van der Zee F, Frijters C, Ersahin M (2015) Celebrating 40 years anaerobic sludge bed reactors for industrial wastewater treatment. Rev Environ Sci Bio/Techno 14:681–702

Vidal G, Soto M, Field J, Mendez PR, Lema JH (1997) Anaerobic biodegradability and toxicity of wastewaters from chlorine and chlorine-free bleaching of eucalyptus kraft pulps. Water Res 31 (10):2487–2494. doi:10.1016/S0043-1354(97)00113-9

Wang Z, Jiang J, Wang X, Fu Y, Li Z, Zhang F, Qin M (2014) Selective removal of phenolic lignin derivatives enables sugars recovery from wood prehydrolysis liquor with remarkable yield. Bioresour Technol 174:198–203. doi:10.1016/j.biortech.2014.10.025

Weiland P (2010) Biogas production: current state and perspectives. Appl Microbiol Biotechnol 85 (4):849–860. doi:10.1007/s00253-009-2246-7

Wen JL, Sun SN, Yuan TQ, Xu F, Sun RC (2013) Fractionation of bamboo culms by autohydrolysis, organosolv delignification and extended delignification: Understanding the fundamental chemistry of the lignin during the integrated process. Bioresour Technol 150:278–286. doi:10.1016/j.biortech.2013.10.015

Wilson DR (2014) www.seai.ie/...Energy.../Waste-to-Energy—Anaerobic-digestion-for-large-industry.p

Yilmaz T, Yuceer A, Basibuyuk M (2008) A comparison of the performance of mesophilic and thermophilic anaerobic filters treating papermill wastewater. Bioresour Technol 99(1):156–163. doi:10.1016/j.biortech.2006.11.038

Zhang A, Shen J, Ni Y (2015) Anaerobic digestion for use in the pulp and paper industry and other sectors: an introductory mini-review. BioResources 10(4):8750–8769

Ziganshin AM, Liebetrau J, Pröter J, Kleinsteuber S (2013) Microbial community structure and dynamics during anaerobic digestion of various agricultural waste materials. Appl Microbiol Biotechnol 97(11):5161–5174. doi:10.1007/s00253-013-4867-0

Chapter 2
Basics of Anaerobic Digestion Process

Abstract Basics of anaerobic digestion process is presented in this chapter. Principal reactions are Hydrolysis, Fermentation Acetogenesis/dehydrogenation, Methanogenesis. The critical step in the anaerobic digestion process is Methanogenesis.

Keywords Anaerobic digestion process · Hydrolysis · Fermentation acetogenesis · Dehydrogenation · Methanogenesis · Acetophilic · Methane bacteria · Hydrogenophilic

In the anaerobic digestion process the organic matter is broken down by a consortium of microorganisms in the absence of oxygen and lead to the formation of digestate and biogas which mainly consist of methane and carbon dioxide. This digestate which is the decomposed substrate resulting from biogas production can be used as a bio-fertilizer (Al Seadi 2001; Kelleher et al. 2000; Chen et al. 2008; Al Seadi et al. 2008). Figure 2.1 shows the anaerobic pathway.

Originally, anaerobic digestion was perceived as a two stage process involving the sequential action of acid forming and methane forming bacteria. Now, it is known to be a complex fermentation process brought about by the symbiotic association of different types of bacteria (Allen and Liu 1998; Edmond-Jacques 1986; Speece 1983; Kosaric and Blaszczyk 1992). The products produced by one group of bacteria serve as the substrates for the next group. The principal reaction sequences can be classified into four major groups involving the following (Fig. 2.1; Table 2.1):

- Hydrolysis
- Fermentation
- Acetogenesis/dehydrogenation
- Methanogenesis

In the first stage i.e. hydrolysis/liquefaction/solubilisation step, large organic polymers such as starches, cellulose, proteins and fats are broken down or

© The Author(s) 2017

P. Bajpai, *Anaerobic Technology in Pulp and Paper Industry*, SpringerBriefs in Applied Sciences and Technology, DOI 10.1007/978-981-10-4130-3_2

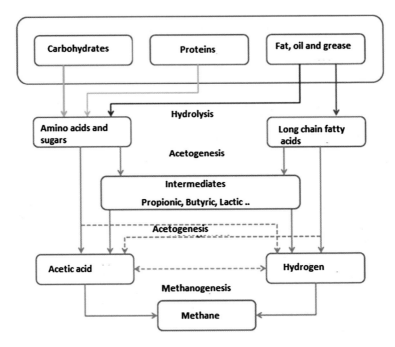

Fig. 2.1 Anaerobic pathway based on Wilson (2014)

Table 2.1 Steps involved in anaerobic oxidation of complex wastes

Hydrolysis
$C_6H_{10}O_4 + 2H_2O \rightarrow C_6H_{12}O_6 + H_2$
Acidogenesis
$C_6H_{12}O_6 \leftrightarrow 2CH_3CH_2OH + 2CO_2$
$C_6H_{12}O_6 + 2H_2 \leftrightarrow 2CH_3CH_2COOH + 2H_2O$
$C_6H_{12}O_6 \rightarrow 3CH_3COOH$
Acetogenesis
$CH_3CH_2COO^- + 3H_2O \leftrightarrow CH_3COO^- + H^+ + HCO_3^- + 3H_2$
$C_6H_{12}O_6 + 2H_2O \leftrightarrow 2CH_3COOH + 2CO_2 + 4H_2$
$CH_3CH_2OH + 2H_2O \leftrightarrow CH_3COO^- + 3H_2 + H^+$
Methanogenesis
$CH_3COOH \rightarrow CH_4 + CO_2$
$CO_2 + 4H_2 \rightarrow CH_4 + 2H_2O$
$2CH_3CH_2OH + CO_2 \rightarrow CH_4 + 2CH_3COOH$
Zupančič and Grilc (2012), Biarnes (2013), Ostrem (2004), Bilitewski et al. (1997), Verma (2002), van Haandel and van der Lubbe (2007), EPA (2006)

depolymerized by acidogenic bacteria into sugars, amino acids, glycerol and long chain fatty acids by hydrolytic exo-enzymes (example, cellulase, amylase, protease, and lipase) excreted by fermentative microorganisms (EPA 2006; van Haandel and van der Lubbe 2007). The hydrolysis reaction is presented in Table 2.1.

The fermentative microorganisms consist of both facultative and strict anaerobes (Broughton 2009). In the enzymatic hydrolysis step, the water-insoluble organics can be solubilized by using water to break the chemical bonds (Parawira 2004) and the resulted simple soluble compounds can be used by the bacterial cells (Gerardi 2003). While some products from hydrolysis such as hydrogen and acetate may be used by the methanogens in the anaerobic digestion process, the majority of the molecules, which were still relatively large, must be further converted to small molecules example acetic acid, so that they may be used to produce methane (Biarnes 2013). Hydrolysis is a relatively slow step and it can limit the rate of the overall anaerobic digestion process, especially when using solidwaste as the substrate (van Haandel and van der Lubbe 2007).

Hydrolysis is immediately followed by the acid-forming step—acidogenesis (Ostrem 2004). In this step, the organics are converted by acid-forming bacteria to higher organic acids such as propionic acid and butyric acid and to acetic acid, hydrogen and carbon dioxide. The higher organic acids are subsequently transferred to acetic acid and hydrogen by acetogenic bacteria. It is always not possible, to draw a clear distinction between acetogenic and acidogenic reactions. Acetate and hydrogen are produced during acidification and acetogenic reactions and both of them are substrates of methanogenic bacteria. The acidogenic and acetogenic bacteria belong to a large and diverse group which includes both facultative and obligate anaerobes. Facultative organisms are able to live in both aerobic and anaerobic environment and obligate are species for which oxygen is toxic. Species isolated from anaerobic digestors include *Clostridium, Peptococcus, Bifidobacterium, Desulfovibrio, Corynebacterium, Lactobacillus, Actinomyces, Staphylococcus, Streptococcus, Micrococcus, Bacillus, Pseudomonas, Selemonas, Veillonella, Sarcina, Desulfobacter, Desulfomonas* and *Escherichia coli* (Kosaric and Blaszczyk 1992). Characteristic of wastewater determine which bacteria predominate.

The hydrogen gas formed in acetogenesis step can be regarded as a waste product of acetogenesis because it inhibits the metabolism of acetogenic bacteria; however, it can be consumed by methane-producing bacteria functioning as hydrogen-scavenging bacteria and converted into methane (Al Seadi et al. 2008).

In the final reaction, methane is produced by methanogenic bacteria. These bacteria are capable of metabolizing formic acid, acetic acid, methanol, carbon monoxide, and carbon dioxide and hydrogen to methane. The methanogenic bacteria are crucial to anaerobic digestion process since they are slow growing and extremely sensitive to the changes in the environment and can assimilate only a narrow array of relatively simple substrates. Some of the notable species that have been classified are *Methanobacterium formicicum, M. bryantic* and *M. thermoautotrophicum; Methanobrevibacter ruminantium, M. arboriphilus* and *M. smithii; Methanococcus vannielli* and *M. voltae;* Methanomicrobium mobile; *Methanogenium cariaci* and *M. marinsnigri, Methanospirilum hungatei* and *Methanosarcina barkei* (Kosaric and Blaszczyk 1992). Two-third of the methane produced during anaerobic microbial conversion is derived from methyl moiety of acetate and about one-third is derived from carbon dioxide reduction. The methanogenic step is the point at which the organic pollution load, in terms of chemical oxygen demand or biochemical oxygen demand is significantly reduced by the anaerobic process since the preceding stages merely convert the organic matter from one form to another. Thus, efficient methanogenesis equates directly with efficient removal of carbonaceous pollution therefore anaerobic wastewater treatment processes have been designed and are operated primarily to satisfy the requirements of this group of bacteria.

Methanogenesis is a critical step in the entire anaerobic digestion process, and its biochemical reactions are the slowest in comparison to those in other steps (Al Seadi et al. 2008). Methane-producing bacteria are strict anaerobes and are vulnerable to even small amounts of oxygen. The methane-producing bacteria can be subdivided into two groups: acetoclastic methane bacteria (acetophilic) and methane bacteria (hydrogenophilic). Another group of methane-producing bacteria is the methyltrophic bacteria which is also able to create methane from methanol (Paul and Liu 2012; Gerardi 2003).

In anaerobic processes where inorganic sulphur is constituent of the wastewater, the sulphate-reducing bacteria—*Desulfovibreo* are also of importance. Sulphate and/or sulphite is present in most effluents from acid sulphite, neutral sulphite semichemical (NSSC), Kraft, chemimechanical (CMP) and chemithermomechanical pulp mills and where aluminum sulphate is used as a sizing agent for paper production. The sulphur-reducing bacteria use sulphate and sulphite as electron acceptors in the metabolism of organic compound to produce hydrogen sulphide and carbon dioxide as end products. Sulphur reduction can become a significant factor in the performance and operation of pulp and paper anaerobic treatment systems. The hydrogen sulphide produced can be both toxic and corrosive. The sulphur reducing and the methane bacteria use and compete for the same organic compounds, reducing methane yield per unit of substrate removed. The methanogenic step is often the most critical one. Disturbance often result in an inhibition or depression of methane formation followed by an excess formation of fatty acids. A small part of the degraded organic matter is converted into new cellulose material. The sludge production rate is low compared with aerobic processes. This means that the sludge retention time must be relatively long if a sufficient amount of biomass is to be obtained in the system. A certain amount of biomass is required for high treatment efficiencies and a stable process.

An obligate, syntrophic relationship exists between the acetogens and methanogens. Synthrophy is the phenomenon that one species lives off the products of another species. The hydrogen partial pressure should be very low so that the thermodynamics become favorable for conversion of volatile acids and alcohols to acetate. Under standard conditions of 1 atm of hydrogen, the free energy change is positive for this conversion and thus precludes it. For example, the free energy change for conversion of propionate to acetate and hydrogen does not become negative until the hydrogen partial pressure decreases below 10^{-4} atm. This relationship has been shown by McCarty (1982). Therefore, it is obligatory that the hydrogen-utilizing methanogens maintain these extremely low hydrogen partial pressures in the system or else, the higher volatile acids, such as propionic and butyric acid, will accumulate in the system. Fortunately, the hydrogen utilizing methanogens in this physiological partnership are adept at this and normally perform this service with ease to allow the reaction to proceed efficiently all the way to methane production. This phenomenon of interspecies hydrogen transfer, which is important to anaerobic biotechnology is a very interesting symbiosis discovered by Brynt et al. (1967).

It is common for the bacterial population concentration to be higher than 1016 cells/ml in case of a well-functioning anaerobic digester (Amani et. al. 2010). This population is typically made of saccharolytic, proteolytic and lipolytic bacteria and methanogens (Gerardi 2003). Of these organisms, the methanogens are known to be highly sensitive to their environment in terms of temperature, pH, and the concentrations of certain chemical compounds (ammonia, volatile fatty acids) (Manser 2015). These are also the slowest growing organisms in the anaerobic digestion reactor. Methanogens are totally dependent on the acetogens and acidogens to survive, as these two organisms convert simple monomers produced during the hydrolysis step into volatile fatty acids and then into acetic acid, carbon dioxide and hydrogen (lipolytic) to supply the methane production process. This relationship is symbiotic as methanogens maintain the digester environment by consuming the protons and volatile fatty acids produced during acidogenesis and acetogenesis, which otherwise would become inhibitory to the biodegradation process.

The supply of hydrogen is often the limiting step in methane production in anaerobic digestion systems (Gerardi 2003). Currently, several research projects are being performed to optimize this aspect of anaerobic digestion system design and operation. Another limiting step in the production of methane is the accumulation of volatile fatty acid in the reactor produced during the acidogenesis step. This balance can be difficult to manage on a large scale because acidogens and acetogens continuously produce compounds that reduce the pH of the system below the preferred range of 6.4–8 for methanogens if sufficient buffering capacity is not available (Speece 1996; Rittmann and McCarty 2001). This type of inconsistency can promote ineffective biogas production in reactors which do not have strict control over the operating environment. Overall, the methanogens sensitivity to the reactor environment also creates an ideal setting for microorganisms, including some pathogens, to survive and possibly multiply during their residence in the system.

References

Al Seadi T (2001) Good practice in quality management of AD residues from biogas production, Report made for the International Energy Agency, Task 24—energy from biological conversion of organic waste, published by IEA Bioenergy and AEA Technology Environment, Oxfordshire, UK

Al Seadi T, Rutz D, Prassl H, Köttner M, Finsterwalder T, Volk S (2008) More about anaerobic digestion (AD). In: Al Seadi T (ed) Biogas handbook. University of Southern Denmark, Esbjerg

Allen DG, Liu HW (1998) Pulp mill effluent remediation. In: Meyers RA (ed) Encyclopedia of environmental analysis and remediation, vol 6. Wiley Interscience Publication, New York, pp 3871–3887

Amani T, Nosrati M, Sreekrishnan TR (2010) Anaerobic digestion from the viewpoint of microbiological, chemical, and operational aspects—a review. Environ Rev 18:255–278

Biarnes M (2013) Biomass to biogas—anaerobic digestion, E Instruments International. Retrieved 11 Oct 2013. URL: http://www.e-inst.com/biomass-to-biogas

Bilitewski B, Härdtle G, Marek K (1997) Waste management. Springer, Berlin

Broughton AD (2009) Hydrolysis and acidogenesis of farm dairy effluent for biogas production at ambient temperatures. M.S. thesis, Massey University, New Zealand

Brynt MP, Wolin EA, Wolin MJ, Wolfe RS (1967) Methanobacillus omelianskii, a symbiotic association of two species of bacteria. Arch Microbiol 59:20–31

Chen Y, Cheng JJ, Creamer KS (2008) Inhibition of anaerobic digestion process: a review. Bioresour Technol 99(10):4044–4064. doi:10.1016/j.biortech.2007.01.057

Edmond-Jacques N (1986) Biomethanation process. In: Rehm HJ, Reed G (eds) Biotechnology, vol 8. VCH, Weinheim, Germany, pp 207–267

EPA (United States Environmental Protection Agency) (2006) Biosolids technology fact sheet: multi-stage anaerobic digestion. Retrieved 11 Oct 2013. URL: http://water.epa.gov/scitech/wastetech/upload/2006_10_16_mtb_multi-stage.pdf

Gerardi MH (2003) Wastewater microbiology series: the microbiology of anaerobic digesters. Wiley, New York

Kelleher BP, Leahy JJ, Henihan AM, O'Dwyer TF, Sutton D, Leahy MJ (2000) Advances in poultry litter disposal technology—a review. Bioresour Technol 83(1):27–36. doi:10.1016/S0960-8524(01)00133-X

Kosaric N, Blaszczyk R (1992) Industrial effluent processing. In: Lederberg J (ed) Encyclopedia of microbiology, vol 2. Academic Press Inc., New York, pp 473–491

Manser ND (2015) Effects of solids retention time and feeding frequency on performance and pathogen fate in semicontinuous mesophilic anaerobic digesters. Graduate Theses and Dissertations

McCarty PL (1982) One hunderd years of anaerobic treatment. In: Hughes PE, Stafford DA, Wheatley BI, Baader W, Lettinga G, Nyns EJ, Verstraete W, Wentworth RL (eds) Anaerobic digestion. Elsevier Biomedical Press BV, Amsterdam, pp 3–22

Ostrem K (2004) Greening waste: anaerobic digestion for treating the organic reaction of municipal solid wastes. M.S. thesis, Columbia University, New York, NY

Paul E, Liu Y (2012) Biological sludge minimization and biomaterials/bioenergy recovery technologies. Wiley, Hoboken, NJ

Parawira W (2004) Anaerobic treatment of agricultural residues and wastewater. PhD dissertation, Lund University, Lund, Sweden

Rittmann BE, McCarty PL (2001) Environmental biotechnology: principles and applications. McGraw-Hill, Boston, MA

Speece RE (1996) Anaerobic biotechnology for industrial wastewater. Archaea Press, Nashville, TN

Speece RE (1983) Anaerobic biotechnology for industrial wastewater treatment. Env Sci Technol 17(9):416–426

van Haandel A, van der Lubbe J (2007) Handbook biological wastewater treatment. Retrieved 13 Oct. URL: http://www.wastewaterhandbook.com/documents/sludge_treatment/831_anaerobic_digestion_theory.pdf

Verma S (2002) Anaerobic digestion of biodegradable organics in municipal solid wastes. M.S. thesis, Columbia University, New York, NY

Wilson DR (2014) www.seai.ie/…Energy…/Waste-to-Energy—Anaerobic-digestion-for-large-industry.p

Zupančič GD, Grilc V (2012) Anaerobic treatment and biogas production from organic waste. In: Kumar S (ed) Management of organic waste. InTech, Croatia, pp 1–28

Chapter 3
Process Parameters Affecting Anaerobic Digestion

Abstract Process parameters affecting anaerobic digestion are presented in this chapter. The important process parameters are: Anaerobic conditions, Temperature, System pH, Volatile fatty acid content and conversion, Availability of micro and trace nutrients, Mixing, Toxicity, Solid retention time, Volatile solids loading rate and Hydraulic retention time.

Keywords Anaerobic digestion · Anaerobic conditions · Temperature · Volatile fatty acid · Mixing · Toxicity · Solid retention time · Volatile solids loading rate · Hydraulic retention time

The important process parameters affecting anaerobic digestion are presented below:

- Anaerobic conditions
- Temperature
- System pH
- Volatile fatty acid content and conversion
- Availability of micro and trace nutrients
- Mixing
- Toxicity

These parameters may overlap each other (Zhang et al. 2015). For example, volatile acid content can be related to the toxicity of the feedstocks and pH of the system.

3.1 Anaerobic Conditions

Most of the important bacteria within the anaerobic system are obligate anaerobes (an organism that lives and grows in the absence of molecular oxygen). Therefore, complete absence of dissolved oxygen is needed for optimum conditions. This is the most basic requirement. It has led to the use of closed reactors in all of the leading developments of high-rate anaerobic processes (McKinney 1983; Bajpai 2000).

P. Bajpai, *Anaerobic Technology in Pulp and Paper Industry*, SpringerBriefs in Applied Sciences and Technology, DOI 10.1007/978-981-10-4130-3_3

3.2   Temperature

Temperature is one of the most important environmental conditions affecting the rate of reaction. Anaerobic processes like other biological processes strongly depend on temperature. The control of temperature is rather critical in this case. The anaerobic process has three known operating temperature ranges viz.

- Psychrophilic (5–15 °C)
- Mesophilic (35–40 °C)
- Thermophilic (50–55 °C)

Figure 3.1 shows the effect of temperature on anaerobic activity. As a rule of thumb for every 10° rise in temperature the rate of reaction doubles.

Common recurring problems associated with anaerobic digesters are maintenance of optimum digester temperature and loss of heating capability. In general, there are two temperature ranges which provide optimum conditions for anaerobic biodegradation: the mesophilic and thermophilic ranges (Verma 2002). The mesophilic temperature is in the range of 30 to 35 °C, usually around 35 °C, whereas the thermophilic temperature ranges from 50 to 60 °C, usually around 55 °C (Gerardi 2003). Thus, at temperatures between 40 and 50 °C, methane-producing bacteria can be inhibited, which results in a decrease in biogas production.

Most of the industrial scale anaerobic digesters which are operating to date have adopted the mesophilic range. Stabilization of the waste is found to be faster at the higher temperature range and therefore thermophilic digesters are smaller in size than those operating in the mesophilic range. The contents of a digester can be heated by pumping them through external heat exchangers and back to the digester. The walls of the digester can be insulated by concrete, cork board or by an air gap plus brick facing or corrugated aluminum facing over rigid insulation. Not much information is available on the quantitative effect of temperature on reaction rate and at present it is generally considered that in the range of 20–55 °C, the reaction rate approximately doubles for every 10 °C increase in temperature. Since the temperature has a significant effect in single species of bacteria, it will therefore

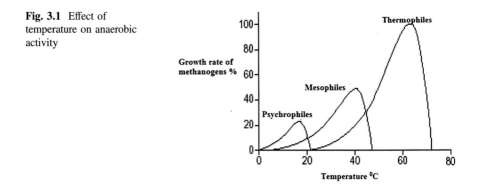

Fig. 3.1 Effect of temperature on anaerobic activity

Table 3.1 Advantages of thermophilic process

Higher rate of biomass hydrolysis in the hydrolysis step
Effective destruction of pathogens
Higher growth rate of methane-producing bacteria at higher temperature and hence higher methane production rate
Reduced retention time, making the process faster and more efficient
Better digestibility and availability of substrates
Better degradation of solid substrates and better utilization of substrate
Better possibility for separating liquid and solid fractions
(Seadi 2008; Nayono 2009)

Table 3.2 Disadvantages of thermophilic process

Large degree of imbalance
Higher energy demand as a result of high temperatures
More sensitivity to toxic inhibitors
Changes in process parameters

have an effect on the species and number of bacteria occurring in a heterogeneous population. Therefore, in the interests of maintaining a stable wastewater treatment process, industrial scale digesters are often temperature controlled to within ±1 °C.

Many modern large anaerobic reactors operate at thermophilic temperature because of its inherent advantages over the mesophilic process which are presented in Table 3.1.

However, the thermophilic process also has its pronounced disadvantages shown in Table 3.2 (Mata-Alvarez 2002; Seadi 2008).

During the digestion process, it is important to keep a constant temperature, as temperature changes or fluctuations will negatively affect the biogas production (Seadi 2008).

3.3 pH and Alkalinity

Two groups of bacteria exist in terms of pH optima namely acidogens and methanogens. The best pH range for acidogens is 5.5–6.5 and for methanogens is 7.8–8.2. The operating pH for combined cultures is 6.8–7.4 with neutral pH being the optimum. Since methanogenesis is considered as a rate limiting step, it is necessary to maintain the reactor pH close to neutral. Low pH reduces the activity of methanogens causing accumulation of volatile fatty acids and hydrogen. At higher partial pressure of hydrogen, propionic acid degrading bacteria will be severely inhibited thereby causing excessive accumulation of higher molecular weight volatile fatty acids such as propionic and butyric acids and the pH drops

Fig. 3.2 Relative activity of
methanogens to pH. Based on
Mata-Alvarez 2002; Seadi
2008

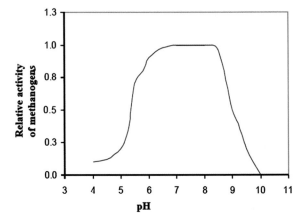

further. If the situation is left uncorrected, the process may eventually fail. This condition is known as a "SOUR" or STUCK" (Fig 3.2).

Joint work of several groups of microorganisms are required for anaerobic digestion of complex organic substrates from which methanogens are the most sensitive to low pH. Below pH 6.5, the growth of methane bacteria slows down and below pH 6.0, the system has serious problems. As the pH increases beyond 7.5, the microbes can grow but the degree of metabolism appears to be reduced. It could be due to the reason that the key nutrients or trace metals are precipitated as the pH increases limiting the metabolism.

Changes in digester operating conditions or introduction of toxic substances can result in process imbalance and also accumulation of volatile fatty acids. Unless the system contains enough alkalinity (buffer capacity), the pH will drop below the optimal levels and the digester will become "sour". Depending on the pH magnitude and the duration of the drop, the production of biogas will reduce to a level where it may completely stop. On the other hand, in a well-operated system, a slight increase of the digester's effluent pH is expected, because organisms produce alkalinity as they consume organic matter rich in protein.

Methanogens are found to be most sensitive to low pH because of the significant inhibiting effect of acidic conditions on their growth (Verma 2002; Lababut and Gooch 2012). Acetogenesis can lead to the formation of organic acids, essentially volatile fatty acids. These acids accounts for the decrease in the system pH. However, maintenance of the system pH in the neutral range from 6.5 to 7.6 is needed for efficient anaerobic digestion (Lababut and Gooch 2012). The methanogenic activity reduces drastically at a pH below 6.3 and above 7.8 and this will inhibit methane production (Leitao et al. 2006). The optimum pH for highest methanogenic activity is in the narrow range of 7.0–7.2 (Ostrem 2004).

For ensuring the health of the methanogens and, thus, continued methane production, pH should be measured throughout the whole process (Biarnes 2013). Lime is used for controlling the pH (Verma 2002). For the most part, the pH is kept at the proper level by the alkalinity which is essentially bicarbonate alkalinity.

Degradation of organic matter results in release of carbon dioxide as an end product. Ammonium ions released from protein degradation react with carbon dioxide to produce ammonium bicarbonate. Salts of organic acids release cation when the acids are metabolized. These cations can also react with carbon dioxide to produce bicarbonates. Alcohols and sugars are neutral compounds that do not have cations to neutralize the acids produced by metabolism and require addition of alkalinity to provide the buffer to maintain the pH at the correct level for good metabolism. Sodium bicarbonate and lime have been used for producing alkalinity in digesters. Under carefully controlled conditions even anhydrous ammonia can be used. The alkalinity must be sufficient to neutralize the volatile acids generated during metabolism. Normal digesters will have 1000–5000 mg/l alkalinity as calcium carbonate. The carbon dioxide in the gas phase will help keep the carbon dioxide concentration in the liquid phase at high levels and will depress the pH unless adequate alkalinity is present. Reducing the gas pressure in the anaerobic system will force a shift in carbon dioxide and will normally increase the pH. Care must be taken in measuring pH of effluent samples from anaerobic systems. If the samples are allowed to sit for more than a few minutes, the excess carbon dioxide will be lost to the atmosphere and the pH will increase, giving the impression that the pH in the anaerobic digester is satisfactory when it may not be satisfactory.

An anaerobic treatment system has its own buffering capacity against pH drop because of alkalinity produced during waste treatment: e.g. the degradation of protein present in the waste releases ammonia which reacts with carbon dioxide forming ammonium carbonate as alkalinity. The degradation of salt of fatty acids may produce some alkalinity. Sulfate and sulfite reduction also produce alkalinity. When pH starts to drop due to volatile fatty acid accumulation, the alkalinity present within the system neutralizes the acid and prevents further reduction in pH. If the alkalinity is not enough to buffer the system pH, we need to add from external as reported earlier.

3.4   Inhibitory Compounds

Methanogenic bacteria in anaerobic systems are mostly sensitive to the presence of toxic compounds, which is the major limitation of anaerobic treatment. These inhibitory substances may be the cause of upset or failure of anaerobic reactor (Chen et al. 2008). These commonly include ammonium, sulfide, light metal ions, heavy metal ions, and some organics. Specifically, the toxic substances may include the following (Gerardi 2003).

- Inorganic sulphur compounds (sulphate, sulphite and sulphide)
- Oxidants including hydrogenperoxide
- Low molecular weight organics
- Heavy metals
- Molecular hydrogen

- Wood constituents including resin acids
- Organic additives such as DTPA

Among them, wood extractives and sulphur compounds are the major toxicants for the anaerobic treatment of pulp mill effluents and their inhibitory effects depend on their concentration. In anaerobic treatment of pulp mill effluents, sulphur inhibition is of most concern and should be reduced.

3.4.1 Sulphur Compounds

Toxicity of inorganic sulphur compounds increases in the following order (Khan and Trottier 1978).

sulphate < thiosulphate < sulphite < sulphide
Sulphate in some pulp and paper industry wastewater has created problems in methanogenic anaerobic treatment. Sulphate reduction occurs also in thermophilic anaerobic reactors (Rintala et al. 1991; Rintala and lettinga 1992). On the other hand anaerobic processes have been used to recover sulphur from sulphate rich pulp and paper industry wastewaters (Särner 1990). In anaerobic treatment, the terminal oxidations are coupled to methane production and dissimilatory reduction of oxidized sulphur compounds present in these wastes. The competition for available substrates such as hydrogen, acetate and methanolby sulphate reducing bacteria and methanogenic bacteria affects the carbon flow. Sulphate reducers also compete with acetogenic bacteria for compounds such as propionate. Thermodynamics and substrate-consumption show an advantage for sulphate reducers over their acetogenic and methanogenic competitors (Thauer 1977; Gottschalk 1983). Hydrogen sulphide production is generally not desirable as it reduces the removal efficiency measured as COD and the methane yield (Ets 1983; Frostell 1984). In addition hydrogen sulphide is toxic, corrosive and contributes to the chemical oxygen demand of the effluent. Undissociated hydrogen sulphide may penetrate cell membranes and therefore is the most toxic sulphide species. Free hydrogen sulphide concentrations of 50 mg/l can inhibit the activity of methanogenic bacteria by about 50% (Kroiss and Wabnegg 1983). Complete inhibition of methanogens has been reported at free sulphide concentrations of approximately 200 mg/l (Lawrence et al. 1984). Later several research groups have shown that anaerobic systems may not be as sensitive to sulphide inhibition as believed earlier (Lettinga et al. 1985; Isa et al. 1986a, b; Koster et al. 1986). Satisfactory operation of anaerobic reactors at organic loading rates of 5–10 kg COD/m³.d has been observed, even in the presence of 200 mg free hydrogen sulphide/l. Sulphide inhibition is most likely to occur with wastewaters that have low COD concentrations and COD sulphate ratio of less than 7.5. In such circumstances, the quantities of biogas produced may be insufficient to strip sulphide from the liquid as it is generated. If the anticipated sulphide levels are excessive, inhibition may be reduced by using the following measures.

- Addition of iron salts to precipitate sulphide from solution
- Control of pH to >8 to reduce free hydrogen sulphide
- Removal of sulphur compounds from the feed
- Hydrogen sulphide stripping and recirculation of the sulphide reactor biogas
- Two stage anaerobic treatment in which sulphur is reduced to hydrogen sulphide and removed in the first stage.

3.4.2 Hydrogen Peroxide

Methanogenic bacteria are strict anaerobes, requiring a highly reduced environment with optimal redox conditions of less than E_C-510 MV. Thus, oxygen and other oxidants present in the feed to anaerobic systems are toxic to methanogens. Hydrogen peroxide frequently used to bleach mechanical pulps is of particular concern. While known to be toxic to bacteria in general, the obligate anaerobic bacteria lack the catalase enzyme necessary for peroxide decomposition. Thus, the methanogenic bacteria are especially sensitive to the presence of hydrogen peroxide. The facultative acidogenic bacteria, however, do produce the catalase enzyme. Physical separation of the acidogenic and methanogenic phases of anaerobic metabolism into two sequential stages is one method of peroxide detoxification (Welander and Anderson 1985). Both the biocatalytic action of the acidogenic bacteria and the chemical reaction with reduced compounds cause peroxide to be decomposed. When anaerobic treatment is followed by activated sludge aerobic polishing, waste activated sludge which also contains facultative acid-forming bacteria can be combined with a hydrogen peroxide-laden effluent in a detoxification pretreatment stage, before single stage anaerobic treatment (McCarty 1982). Other method for removing hydrogen peroxide include decomposition by chemical reaction with reduced compound such as sulphite and sulphide.

3.4.3 Low-Molecular-Weight-Organic Compounds

Low-molecular-weight-organic compounds such as volatile fatty acids, sugars, alcohols are produced in large quantities during pulping. They can be inhibitory to the digestion process which can lead to system failure. Volatie fatty acids encompass a group of following six compounds.

- Acetic acid/acetate
- Propionic acid/propionate
- Butyric acid/butyrate
- Valeric acid/valerate
- Caproic acid/caproate
- Enanthic acid/enanthate

Among these acetate is predominant. In the digesters which are correctly designed and well-operated, the concentration of total volatile fatty acids is typically below 500 mg/l as acetic acid. However, if the digester is undersized for the organic load this concentration can be higher. At volatile fatty acids concentrations over 1500–2000 mg/l, biogas production might be limited by inhibition. However, rather than a specific concentration, it is a sudden and steady increase of volatile fatty acids in the effluent what can be a sign of a digester upset. Thus, it is important to monitor volatile fatty acids periodically in order to detect problems on time, and make the necessary operational changes before digester failure occurs.

3.4.4 Heavy Metals

Heavy metals, are known to be toxic to anaerobic processes by reacting with enzymes to block metabolism. These are generally not a concern in anaerobic treatment of pulp and paper effluents because they precipitate in the presence of sulphide. Iron and nickel, in fact, are two metals that frequently must be added to satisfy micronutrient demand (Pohland 1992).

3.4.5 Molecular Hydrogen

Molecular hydrogen also maybe the most sensitive parameter of process upsets. The energy available for the degradation of propionate is very small, and requires partial pressures of hydrogen below 10–4 atm at 25 °C (McCarty and Smith 1986; Schmidt and Ahring 1993). Such low hydrogen partial pressures in anaerobic digester systems are only possible by the syntrophic relationships between hydrogen-producing bacteria to hydrogen-oxidizing methanogens (Bryant 1979). The balance between these two groups of organisms is of foremost importance for preventing digester upsets (Demirel and Yenigün 2002). As opposed to other parameters, molecular hydrogen is more difficult to measure due to the low levels found in anaerobic digester systems, and requires specialized equipment to determine it.

3.4.6 Wood Constituents

High-molar-mass lignin is recalcitrant towards anaerobic degradation (Zeikus et al. 1982; Hackett et al. 1977; Odier and Monties 1983). However, some studies show slow but detectable anaerobic biodegradation of lignin to methane and carbon dioxide in sediments and thermophilic laboratory conditions (Benner and Hodson

1985; Benner et al. 1984). Anaerobic methanogenic mixed cultures can decompose monomeric, dimeric and oligomeric lignin substructure model compounds (Colberg and Young 1985; Kaiser and Hanselmann 1982; Grbic-Galic 1983; Chen et al. 1985). The relationship between lignin polymer length and its anaerobic biodegradability has been reviewed by Field (1989). The share of polymeric and oligomeric lignin compounds in the pulp mill wastewater COD mainly determines its recalcitrance towards anaerobic degradation. Wastewater lignins are either nontoxic or may show some inhibition towards methanogenesis at high concentrations (3300–6000 mg/l) (Sierra-Alvarez and Lettinga 1991). Cellulose is easily degradable in anaerobic systems. The degradation of cellulose and hemicellulose decreases when the polysaccharides are complexed with lignin (Benner and Hodson 1985; Benner et al. 1984). Resin acids may also inhibit anaerobic treatment of wastes containing these compounds at high concentrations (Welander et al. 1988). The hydrolyzable tannin, gallotannic acid, is toxic to methanogens (Field et al. 1988). The role of natural wood constituents on the anaerobic treatability of forest industry wastewaters have been studied by Sierra Alvarez and Lettinga (1990). Several resin compounds including volatile terpenes, apolar phenols, resin acids were studied for methanogenic toxicityand were shown to cause 50% inhibition in concentrations ranging from 20 to 330 mg/l.

3.4.7   DTPA

DTPA used for stabilizing hydrogen peroxide in bleaching of mechanical pulp have been reported to be inhibitory or toxic to anaerobic organisms (Welander and Anderson 1985). These organic compounds have been detoxified by precipitation with aluminium, iron and calcium salts.

3.5   Nutrients and Trace Elements

All microbial processes including anaerobic process requires macro (N, P and S) and micro (trace metals) nutrients in sufficient concentration to support biomass synthesis. In addition to N and P, anaerobic microorganisms especially methanogens have specific requirements of trace metals such as Ni, Co, Fe, Mo, Se etc. The nutrients and trace metals requirements for anaerobic process are much lower as only 4–10% of the COD removed is converted biomass. Significant differences in nutrient requirements are found between aerobes and anaerobes (Suryawanshi et al. 2013). Differences in critical nutrients emanate due to unique enzyme systems required by methane-forming bacteria. In the conversion of acetate to methane, cobalt, iron, nickel, sulfur, selenium, tungsten and molybdenum are required. Additional micronutrients are barium, calcium, magnesium and sodium.

Macronutrient requirements for anaerobic processes are much lower than the requirements for aerobic processes due to lower cell yield. Nitrogen and phosphorus are made available to anaerobic processes as ammonical- nitrogen and orthophosphate. Their amount needed to satisfy acceptable digester performance can be determined, considering adequate residual concentrations of soluble nutrients in the digester effluent. Residual values of 5 mg/l of NH4 + and 1–2 mg/l of HPO4–are usually recommended. Absence of residual nutrients means that nutrients must be added. While for nitrogen addition, ammonium chloride, aqueous ammonia and urea may be used, for phosphorus addition, phosphate salts and phosphoric acid may be used. While some methanogens are able to fix nitrogen, some use the amino acid, alanine. Nutrient requirements for anaerobic digesters vary as a function of OLR. Generally, COD: nitrogen: phosphorous of 1000:7:1 is used for high strength wastes and 350:7:1 for low strength wastes, respectively. The carbon/nitrogen value of at least 25:1 is suggested for optimal gas production. Nitrogen is approximately 12% and phosphorus 2% of the dry weight of bacterial cells. Both nitrogen and phosphorous should not be limited in the digester.

3.6 Solids Retention Time (SRT)

One of the most important requirement in the design and operation of anaerobic systems for the treatment of largely soluble industrial wastewaters is that the essential anaerobic bacteria should not be washed out of the system at a greater rate than they can multiply (Hall 1992). The average length of time with each cell remains within the treatment system is termed as the mean solids retention time (SRT). It is also known as mean cell residence time. The relationship between the reactor volume and volumetric flow rate is mostly used to define the SRT of a completely mixed system. In the anaerobic digestion system, SRT is one of an important operating factor to consider because the consumption of the substrate is controlled by the kinetics of the microorganisms. SRT is calculated as the total mass of bacteria within the digester divided by the total mass of bacteria lost from the system in unit time.

$$\text{SRT} = \frac{\text{Mass of solids in digester}(\text{kg})}{\text{Rate of removal of solids from digester}(\text{kg/day})}$$

For maximizing the removal capacity of the digester, the SRT is maintained at the highest possible value. In addition to reducing the required volume of digester, high SRT systems provide significant buffering capacity for protection against the effects of shock loadings and toxic or inhibitory substances in the feed (Hall 1992). During a period of inhibition, the prevailing microorganism growth rate, μ, is forced to decrease and the minimum SRT required to accommodate this growth rate increases. If normally operated at a low SRT, the system may approach biomass growth-limited conditions after exposure to toxicants. Operation at a high SRT affords a safety factor to protect against system failure and also to allow biological acclimation to the inhibiting material. It is provision of a long mean SRT (greater than 30 days) which has led to the development of a wide range of anaerobic processes, differing essentially only in their method of retaining the active bacteria (Hall 1992).

The optimal SRT finally will be a function of the waste composition, operating temperature, type of reactor, and other process details (Buekens 2005). In general it can be presumed that a longer SRT allows for more degradation and pathogen inactivation of the substrate when compared to a shorter SRT under similar operating conditions. The SRT is a very important design parameter to work with because the bacteria and archaea providing the carbon conversion have an optimum time that they need to be in the reactor for to perform their metabolism and produce methane. If SRT is insufficient, the microorganisms will wash out of the reactor. According to Dohányos et al. (2001), a rule of thumb often followed in the design of anaerobic digestion system is the use of SRT that is at least two times the generation time of the methanogens under the digester conditions. Weimer (1998) reported that slow-growing mesophilic methanogens requires up to a 130 h generation time, which correlates to 5 days or a minimum SRT of 10 days.

In terms of pathogen reduction, the SRT can be an important ally for the operator. Chen et al. (2012) reported that a completely-mixed mesophilic anaerobic digestion process was able to remove *E. coli* and *Salmonella sp.* from the influent with removal efficiencies of 1.93, 2.98 and 3.01 log10 units for E. coli and 1.93, 2.76 to 3.72 log10 units for *Salmonella sp.* This improvement took place as the SRT was increased from 11 days to 16 days to 25 days and highlights an aspect of the reactor that can be optimized for killing the pathogens. The difficulty with SRT and pathogen removal is that SRT represents an average cell residence time, which means that there will be a percentage of cells that are in the digester for periods which are both longer and shorter as compared with this value. If the cell is pathogenic and has a short enough residence time, it may come out from the digester in a viable state. This is dependent on the operating parameters of the digester and the inactivation characteristics of the cell.

3.7 Volatile Solids Loading Rate

The amount of volatile solids requiring digestion is often used as the basis of design in anaerobic digestion (Bajpai 2000). Volatile solids are defined as those solids (mostly organic) which are oxidized and driven off as gas at 500±50 °C. Specifically the mass (kg) of volatile solids added per day per cubic meter of digester capacity, or the mass of volatile solids added per day per kilogram of volatile solids in the digester, are used as criteria, that is.

$$\text{Volatile solids loading rate} = \frac{\text{Volatile solids added per day} (\text{kg}/\text{day})}{\text{Volume of digester} (\text{m}^3)}$$

or

$$\text{Volatile solids loading rate} = \frac{\text{Volatile solids added per day} (\text{kg}/\text{day})}{\text{Mass of volatile solids in digester} (\text{kg})}$$

3.8   Hydraulic Retention Time

The hydraulic retention time (HRT, in days) is the average retention time of the wastewater in the digester (Bajpai 2000). It can be calculated as.

$$\text{HRT} = \frac{V}{Q}$$

where V = volume of digester (m^3)
and Q = flow rate of wastewater through the digester (m^3/day)
Minimal HRT reduces the reactor volume and thus reduces capital cost.

3.9   Mixing

Reactor mixing is an important operational characteristic for anaerobic digesters. Three mixing strategies are being used in anaerobic digestion systems: continuous, intermittent and minimal. Insufficient mixing leads to following problems (Kaparaju et al. 2008; Karim et al. 2005):

– Non-uniform distribution of substrates, enzymes and microorganisms
– Incomplete stabilization of the waste
– A reduction in methane production and pathogen destruction.

Kinyua (2015) reported that unmixed digesters in Costa Rica showed adequate biogas production and effluent quality. There are conflicting opinions about which method is best in terms of biogas production. Chen et al. (1990) have suggested that very small mixing is required to promote the symbiotic lifestyle between the methanogens and the acetogens, which is improved by their close spatial proximity to each other. This may be disrupted by over-mixing. This can also damage the cell walls of the microorganisms (Kaparaju et al. 2008). On the other side, continuous mixing has been shown to increase biogas production when compared to unmixed cases (Karim et al. 2005; Ho et al. 1985). More research is needed to better understand how anaerobic systems that are unmixed perform compared to the mixed systems, because eliminating the requirement to mix will reduce the energy demand of the anaerobic digester.

References

Bajpai P (2000) Anaerobic treatment of pulp and paper industry effluents. Pira Technology Series, UK

Benner R, Hodson RE (1985) Thermophilic anaerobic biodegradation of [^{14}C] lignin, [^{14}C] cellulose [^{14}C] lignocellulose preparations. Appl Environ Microbiol 1985(50):971–976

Benner R, MacCubbin AE, Hodson RE (1984) Anaerobic biodegradation of the lignin and polysaccharide components of lignocellulose and synthetic lignin by sediment microflora. Appl Environ Microbiol 1984(47):998–1004

Biarnes M (2013) Biomass to biogas—Anaerobic digestion. E Instruments International, http://www.e-inst.com/biomass-to-biogas

Bryant MP (1979) Microbial Methane production—theoretical aspects. J Anim Sci 48(1):193–201

Buekens A (2005) Energy recovery from residual waste by means of anaerobic digestion technologies. In: Proceedings of the future of residual waste management in Europe. November Luxembourg, 17–18, 2005

Chen T, Fu B, Wang Y, Jiang Q, Liu H (2012) Reactor performance and bacterial pathogen removal in response to sludge retention time in a mesophilic anaerobic digester treating sewage sludge. Bioresour Technol 106:20–26

Chen TH, Chynoweth P, Biljetina R (1990) Anaerobic digestion of municipal solid waste in a non-mixed solids concentrating digester. Applied Biochemical Biotechnology 24–25(1):533–544

Chen W, Supanwong K, Ohmiya K, Shimazu S, Kawakami H (1985) Anaerobic biodegradation of veratrylglycerol-beta-guaiacyl ether and guaiacoxylacetic acid by mixed rumen bacteria. Appl Environ Microbiol 50:1451–1456

Chen Y, Cheng JJ, Creamer KS (2008) Inhibition of anaerobic digestion process: A review. Bioresour Technol 99(10):4044–4064. doi:10.1016/j.biortech.2007.01.057

Colberg PJ, Young LY (1985) Anaerobic degradation of soluble fractions of (14Lignin)-lignocellulose. Appl Environ Microbiol 49:345–349

Demirel B, Yenigün O (2002) Two-phase anaerobic digestion processes: a review. J Chem Technol Biotechnol 77(7):743–755

Doháynos M, Zábranská J (2001) Sludge into biosolids: processing, disposal, and utilization. IWA Publishing, London

Ets BJ, Ferguson JF, Benjamin MM (1983) The fate and effect of bisulphite in anaerobic treatment. J Water Pollution Control Fed 1983(55):1355–1365

Field JA, Leyendeckers MJH, Sierra-Alvarez R, Lettinga G, Habets LHA (1988) The methanogenic toxicity of bark tannins and the anaerobic biodegradability of water soluble bark matter. Water Sci Technol 20(1):219–240

Field JA (1989) The effect of tannic compounds on anaerobic wastewater treatment. Doctoral thesis, Dept. Water Pollution Control, Agricultural University of Wageningen, The Netherlands

Frostell B (1984) Anaerobic-aerobic pilot-scale treatment of a sulphite evaporator condensate. Pulp Paper Canada 1984(85):80–87

Gerardi MH (2003) The microbiology of anaerobic digesters. Wiley, Hoboken, NJ

Gottschalk G (1983) Bacterial metabolism, 2nd edn. Springer Verlag, New York 1983

Grbic-Galic D (1983) Anaerobic degradation of coniferyl alcohol by methanogenic consortia. Appl Environ Microbiol 46:1442–1446

Hackett WF, Connors WJ, Kirk TK, Zeikus JG (1977) Microbial decomposition of synthetic ^{14}C-labeled lignins in nature: Lignin biodegradation in a variety of natural materials. Appl Environ Microbiol 1977(33):43–51

Hall ER (1992) Anaerobic treatment of wastewaters in suspended growth and fixed film processes. In: Malina Jr. JF, Pohland FG (eds) Design of anaerobic processes for the treatment of industrial and municipal wastes, Technomic Publishing Comp. Inc., Lancester, USA, pp 41–110

Ho CC, Tan YK (1985) Anaerobic treatment of palm oil mill effluent by tank digesters. J Chem Technol Biotechnol 35(2):155–164

Isa Z, Grusenmeyer S, Verstraete W (1986a) Sulfate reduction relative to methane production in high rate anaerobic digestion: Technical aspects. Appl Env Microbiol 51(3):572–579

Isa Z, Grusenmeyer S, Verstraete W (1986b) Sulfate reduction relative to methane production in high rate anaerobic digestion:microbiological aspects. Appl Env Microbiol 51(3):580–587

Kaiser JP, Hanselmann KW (1982) Aromatic chemicals through anaerobic microbial conversion of lignin monomers. Experientia 1982(38):167–176

Kaparaju P, Buendia I, Ellegaard L, Angelidakia I (2008) Effects of mixing on methane production during thermophilic anaerobic digestion of manure: lab-scale and pilot-scale studies. Bioresour Technol 99(11):4919–4928

Karim K, Hoffmann R, Thomas Klasson K, Al-Dahhan MH (2005) Anaerobic digestion of animal waste: effect of mode of mixing. Water Res 39(15):3597–3606

Khan AW, Trottier TM (1978) Effect of sulfur containing compounds on anaerobic degradation of cellulose to methane by mixed cultures obtained from sewage sludge. Appl Env Microbiol 35:1027–1034

Kinyua, MN (2015) Energy production and effluent quality in tubular digesters treating livestock waste in Rural Costa Rica. Graduate Theses and Dissertations. http://scholarcommons.usf.edu/etd/5716

Koster IW, Rinzeman A, de Vegt L, Lettinga G (1986) Sulfide inhibition of the methanogenic activity of granular sludge at various pH levels. Water Res 1986(24):313–319

Kroiss H, Wabnegg FP (1983) Sulphide toxicity with anaerobic wastewater treatment. In: Proceeding of european symposium on anaerobic wastewater treatment (AWWT), TNO Corporate Communications Dept., The Hague, The Netherlands

Labatut RA, Gooch CA (2012) Monitoring of anaerobic digestion process to optimize performance and prevent system failure, retrieved Oct 17 2013. URL: http://www.abc.cornell.edu/prodairy/gotmanure/2012proceedings/21.Rodrigo.Labatut.pdf

Lawrence AW, McCarty PL, Guerin A (1964) The effects of sulfides on anaerobic treatment. Proc 19th Indust Waste Conf Purdue University 1964:343

Leitao RC, van Haandel AC, Zeeman G, Lettinga G (2006) The effects of operational and environmental variations on anaerobic wastewater treatment systems: a review. Bioresour Technol 97(9):1105–1118. doi:10.1016/j.biortech.2004.12.007

Lettinga G, Zeeuw W de, Hulshoff Pol L (1985) Anaerobic wastewater treatment based on biomass retention with emphasis on the UASB Process. In: Proceeding of 4th international symposium on anaerobic digestion, Guangzhou, People's Republic of China, China State Biogass Association

Mata-Alvarez J (2002) Fundamentals of the anaerobic digestion process. IWA Publishing Company, UK

McCarty PL (1982) One Hunderd years of anaerobic treatment. In: Hughes PE, Stafford DA, Wheatley BI, Baader W, Lettinga G, Nyns EJ, Verstraete W, Wentworth RL (eds) Anaerobic Digestion, Elsevier Biomedical Press BV, Amsterdam, pp 3–22

McCarty PL, Smith DP (1986) Anaerobic Waste-Water Treatment. 4. Environ Sci Technol 20 (12):1200–1206

McKinney RE (1983) Anaerobic treatment concepts. Proc Tappi 1983 Environ Confer Tappi 1983:163–172

Nayono SE (2009) Anaerobic digestion of organic solid waste for energy production. KIT Scientific Publishing, Germany

Odier E, Monties B (1983) Absence of microbial mineralization of lignin in anaerobic enrichment cultures. Appl Environ Microbiol 1983(46):661–665

Ostrem K (2004) Greening waste: anaerobic digestion for treating the organic reaction of municipal solid wastes. M.S. thesis, Columbia University, New York, NY

Pohland FG (1992) Anaerobic treatment: fundamental concepts, applications and new horizons. In: Malina Jr. JF, Pohland FG, (eds) Design of anaerobic processes for the treatment of industrial and municipal wastes, Technomic Publishing Comp. Inc. Lancester, USA, pp 1–33

Rintala J, Lettinga G (1992) Effect of temperature elevation from 37 to 55 °C on anaerobic treatament of sulphate rich acidified wastewaters. Env Technol 1992(13):810–812

Rintala J, Sanz Martin JL, Lettinga G (1991) Thermophilic anaerobic treatment of sulfate-rich pulp and paper integrate process water. Wat Sci Technol 1991(24):149–160

Särner E (1990) Removal of sulphate and sulphite in an anaerobic trickling (ANTRIC) filter. Water Sci Technol 22(1–2):395–404

Schmidt JE, Ahring BK (1993) Effects of hydrogen and formate on the degradation of propionate and butyrate in thermophilic granules from an upflow anaerobic sludge blanket reactor. Appl Environ Microbiol 59(8):2546–2551

Seadi TA (2008) Biogas handbook: http://www.lemvigbiogas.com/BiogasHandbook.pdf

Sierra-Alvarez R, Lettinga G (1990) The methanogenic toxicity of wood resin constituents. Biol Wastes 33:211–226

Sierra-Alvarez R, Lettinga G (1991) The methanogenic toxicity of wastewater lignins and lignin related compounds. J Chem Technol Biotechnol 50:443–455

Suryawanshi PC, Chaudhari B, Bhardwaj S, Yeole TY (2013) Operating procedures for efficient anaerobic digester operation. Res J Anim Vet Fish Sci 1:12–15

Thauer RK, Jungermann K, Decker K (1977) Energy conservation in chemotrophic anaerobic bacteria. Bacteriol Rev 1977(41):100–180

Verma S (2002) Anaerobic digestion of biodegradable organics in municipal solid wastes. M.S. thesis, Columbia University, New York, NY

Weimer PJ (1998) Manipulating ruminal fermentation: a microbial ecological perspective. J Anim Sci 76(12):3114–3122

Welander T, Anderson PE (1985) Anaerobic treatment of wastewater from the production of chemithermomechanical pulp. Water Sci Technol 17(1):103–112

Welander T, Malmqvist A, Yu P (1988) Anaerobic treatment of toxic forest industry wastewaters. In: Hall ER, Hobson PN, (eds) Anaerobic Digestion, Pergamon Press, New York, pp 267–274

Zeikus JG, Wellstein AL, Kirk TK (1982) Molecular basis for the biodegradative recalcitrance of lignin in anaerobic environments. FEMS Microbiol Lett 1982(15):193–197

Zhang A, Shen J, Ni Y (2015) Anaerobic digestion for use in the pulp and paper industry and other sectors: an introductory mini-review. BioResources 10(4):8750–8769

Chapter 4
Comparison of Aerobic Treatment with Anaerobic Treatment

Abstract Comparison of anaerobic treatment with aerobic treatment is presented in this chapter. Aerobic processes take place in the presence of air and anaerobic treatment processes takes place in the absence of air. The final products of organic assimilation in anaerobic Waste Water Treatment Plant are methane, carbon dioxide gas and biomass. The anaerobic treatment is now becoming a viable alternative due to its general advantages over aerobic processes.

Keywords Anaerobic treatment · Aerobic treatment · Organic assimilation · Waste water treatment plant · Methane · Carbon dioxide · Biomass

Aerobic processes take place in the presence of air and utilize those microorganisms (also called aerobes), which use molecular/free oxygen to assimilate organic impurities and convert them into carbon dioxide, water and biomass. The anaerobic treatment processes, on other hand take place in the absence of air (and thus molecular/free oxygen) by those microorganisms (also called anaerobes) which do not require air (molecular/free oxygen) to assimilate organic impurities. The final products of organic assimilation in anaerobic Waste Water Treatment Plant are methane and carbon dioxide gas and biomass. Aerobic biological treatment has been used for a long time to reduce the amount of organic pollutants in pulp and paper mill effluents and extensive experience of this method is available. The anaerobic treatment is now becoming a viable alternative due to its general advantages over aerobic processes as shown in Table 4.1 (Rintala and Puhakka 1994; Speece 1983; Lee 1993; Allen and Liu 1998). Anaerobic treatment methods can lead to the application of integrated environmental protection systems. In principal, they can be combined with post treatment methods by which useful bulk products like ammonia or sulphur can be recovered, while in specific cases, effluents and excess sludge could be employed for irrigation and fertilizers or soil conditioning. Anaerobic treatment has also some drawbacks (Table 4.2) (Rintala and Puhakka 1994; Speece 1983; Lee 1993; Allen and Liu 1998). Specific limitations of anaerobic treatment are presented in the below paragraphs:

© The Author(s) 2017
P. Bajpai, *Anaerobic Technology in Pulp and Paper Industry*, SpringerBriefs in Applied Sciences and Technology, DOI 10.1007/978-981-10-4130-3_4

Table 4.1 Advantages of anaerobic process

Less energy requirement as no aeration is needed 0.5–0.75 kWh energy is needed for every 1 kg of COD removal by aerobic process
Energy generation in the form of methane gas 1.16 kWh energy is produced for every 1 kg of COD removal by anaerobic process
Less biomass generation Anaerobic process produces only 20% of sludge that of aerobic process
Less nutrients requirement Lower biomass synthesis rate also implies less nutrients requirement: 20% of aerobic
Application of higher organic loading rate Organic loading rates of 5–10 times higher than that of aerobic processes are possible
Space saving Application of higher loading rate requires smaller reactor volume thereby saving the land requirement
Ability to transform several hazardous solvents including chloroform, trichloroethylene and trichloroethane to an easily degradable form

The effluents from pulp and paper industry contain several types of phenolic compounds ranging from simple monomers to high molecular weight polyphenolic polymers. The low molecular weight phenolics are generally biodegradable including lignin derived monomers and chlorinated phenolics (Colberg 1988; Schink 1988). As the molecular weight of phenolic compounds increases, a sharp decrease in their anaerobic biodegradability is seen. High molecular weight lignin and tannins are not biodegradable in anaerobic environments (Zeikus et al. 1982; Colberg and Young 1985). Wastewaters such as black liquors and bleaching effluents, are generally only 50% biodegradable or less (Sierra-Alvarez et al. 1991; Qui et al. 1988, 34, 35). In these wastewaters, lignin can account for 50% of the COD. Semi-chemical and chemithermomechanical pulping liquors also contain significant amount of lignin and are thus not fully biodegradable (Wilson et al. 1987; Welander and Anderson 1985; Jurgensen et al. 1985). No significant colour removal can be expected by anaerobic treatment since colour is an important characteristics associated with high molecular weight polyphenolic and lignin. Anaerobic treatment is also inhibited by the presence of toxic substances which can interfere with the metabolism of readily biodegradable substances. Common toxic organic substances in pulp and paper industry effluents include: resin compounds, chlorinated phenolics and tannins (Salkinoja-Salonen et al. 1984; Guthrie et al. 1984; Field et al. 1988). Resinous components of wood, such as resin acids and volatile terpenes are important since they are present in many types of effluents produced from industrial processes involving alkaline treatments of wood. Resin acids and volatile terpenes cause methanogenic inhibition even at low concentrations. Low molecular weight chlorinated phenols, which are present in low

Table 4.2 Limitations of anaerobic processes

Long start-up time Because of lower biomass synthesis rate, it requires longer start-up time to attain a biomass concentration
Long recovery time If an anaerobic system subjected to disturbances either due to biomass wash-out, toxic substances or shock loading, it may take longer time for the system to return to normal operating condition
Specific nutrients/trace metal requirements Anaerobic microorganisms especially methanogens have specific nutrients e.g. Fe, Ni, and Co requirement for optimum growth
More susceptible to changes in environmental conditions Anaerobic microorganisms especially methanogens are prone to changes in conditions such as temperature, pH, redox potential, etc.
Treatment of sulfate rich wastewater The presence of sulfate not only reduces the methane yield due to substrate competition but also inhibits the methanogens due to sulfide production
Effluent quality of treated wastewater The minimum substrate concentration from which microorganisms are able to generate energy for their growth and maintenance is much higher for anaerobic treatment system. Owing to this fact, anaerobic processes may not able to degrade the organic matter to the level meeting the discharge limits for ultimate disposal
Treatment of high protein and nitrogen containing wastewater The anaerobic degradation of proteins produces amines which are no longer be degraded anaerobically. Similarly nitrogen remains unchanged during anaerobic treatment Recently, a process called ANAMMOX (Anaerobic Ammonium Oxidation) has been developed to anaerobically oxidize NH_4^+ to nitrogen in presence of nitrite

concentrations in bleaching effluents are potentially toxic to anaerobic digestion processes. These compounds are highly toxic to methane bacteria at very low concentrations. Generally, the methanogenic toxicity of chlorinated phenols increases with increasing Cl-number and also with increasing apolarity. Tannic compounds are less toxic than resin compounds and chlorinated phenols but they are present at fairly high concentrations in debarking wastewater and in fibre board effluents (Field et al. 1989). The organic toxins not only present problems for anaerobic digestion processes, but they are also known to cause toxicity to the aquatic organisms of the discharge environment. The toxicity to fish has been shown for resin compounds, chlorinated phenols and tannins (Roger 1973; Leach and Thakore 1976; Kaser et al. 1984; Junna et al. 1982; Temmink et al. 1989). The anaerobic treatment systems have the limited capacity to decrease the aquatic toxicity of forest industry wastewaters (Wilson et al. 1987). Resin compounds are poorly degraded by anaerobic microorganisms (Schink 1985). Low molecular

Table 4.3 Comparison between aerobic and anaerobic treatment processes

Anaerobic	Aerobic
Organic loading rate	
High loading rates: 10–40 kg COD/m^3-day (for high rate reactors, e.g. AF, UASB, E/FBR)	Low loading rates: 0.5–1.5 kg COD/m^3-day (for activated sludge process)
Biomass yield	
Low biomass yield: 0.05–0.15 kg VSS/kg COD (biomass yield is not constant but depends on types of substrates metabolized)	High biomass yield: 0.35–0.45 kg VSS/kg COD (biomass yield is fairly constant irrespective of types of substrates metabolized)
Specific substrate utilization rate	
High rate: 0.75–1.5 kg COD/kg VSS-day	Low rate: 0.15–0.75 kg COD/kg VSS-day
Start-up time	
Long start-up: 1–2 months for mesophilic: 2–3 months for thermophilic	Short start-up: 1–2 weeks
SRT	
Longer SRT is essential to retain the slow growing methanogens within the reactor	SRT of 4–10 days is enough in case of activated sludge process
Microbiology	
Anaerobic process is multi-step process and diverse group of microorganisms degrade the organic matter in a sequential order	Aerobic process is mainly a one-species phenomenon
Environmental factors	
The process is highly susceptible to changes in environmental conditions	The process is less susceptible to changes in environmental conditions
Carbon balance	
About 95% to biogas and 5% to biomass	About 50% to biomass and 50% to CO_2
Energy balance	
90% recovered as biogas, 5–7% for cell growth and 3–5% wasted as heat	About 60% stored in new cells and 40% lost as process heat
Electricity consumption per metric ton COD destroyed	1100 kWh
Methane generation per metric ton COD destroyed 1.16 × 10^7 kJ	
Net cell production per metric ton COD destroyed 20–150 kg	400–600 kg

weight tannins of bark are only partially degraded during anaerobic treatment (Field 1989). On the other hand monomeric chlorinated phenols are highly metabolized during anaerobic treatment which can result in significant aquatic toxicity removal (Wood et al. 1989; Mikesell and Boyd 1986; Tiedje et al. 1987; Hakulinen and Salkinoja-Salonen et al. 1982a, b). Pulp and paper manufacturing use sulphur in various forms and processes. Sulphur compounds in wastewaters may inhibit the

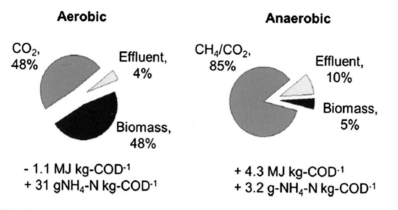

Fig. 4.1 Generalized comparison between aerobic and anaerobic wastewater treatment in terms of the fate of organic carbon [expressed as chemical oxygen demand (COD)] and energy production/consumption and nutrient requirements (expressed as N-requirements) [adapted from van Lier et al. (2008)]

methane producing bacteria or act as terminal electron acceptors for sulphate reducing bacteria which may compete for the available substrates. This may result in low loading rate potentials and shift the population to SRB instead of MPB (Lettinga et al. 1991).

The comparison between aerobic and anaerobic microbial degradation processes is shown in Table 4.3 and Fig. 4.1. The major factors for comparison are electrical power usage, methane gas production and excess microbial cell production which has an associated disposal cost. The comparison shown in Table 4.3 is based on a ton of chemical oxygen demand (COD) destroyed. McDermott (1983) has reported that the net operating cost differential between anaerobic and aerobic treatment is approximately $160 per metric ton less for the anaerobic process (assuming $0.06/kWh, $4.50/$10^6$ Btu for methane and $100/ton of dry cell mass disposal costs). For some industries, this cost differential may be as high as $250.

References

Allen DG, Liu HW (1998) Pulp mill effluent remediation. In: Meyers RA (ed) Encyclopedia of environmental analysis and remediation, vol 6. Wiley, Wiley Interscience Publication, New York, pp 3871–3887

Colberg PJ (1988) Anaerobic microbial degradation of cellulose, lignin, oligolignols and monoaromatic lignin derivatives. In: Zehnder AJB (ed) Biology of anaerobic microorganisms. Wiley, New York, pp 333–342

Colberg PJ, Young LY (1985) Anaerobic degradation of soluble fractions of ([14]Lignin)-lignocellulose. Appl Environ Microbiol 49:345–349

Field JA (1989) The effect of tannic compounds on anaerobic wastewater treatment. Doctoral thesis, Dept. Water Pollution Control, Agricultural University of Wageningen, Wageningen, The Netherlands

Field JA, Leyendeckers MJH, Sierra-Alvarez R, Lettinga G, Habets LHA (1988) The methanogenic toxicity of bark tannins and the anaerobic biodegradability of water soluble bark matter. Water Sci Technol 20(1):219–240

Field JA, Kortekaas S, Lettinga G (1989) The tannin theory of methanogenic toxicity. Biol Wastes 29:241–262

Guthrie MA, Kirsch EJ, Wukasch RF, Grady CPL Jr (1984) Pentachlorophenol biodegradation-II. Water Res 18:451–461

Hakulinen R, Salkinoja-Salonen M (1982a) Treatment of Kraft bleaching effluents: comparison of results obtained by Enso-Fenox and alternative methods. In: International pulp bleaching conference, pp 97–106

Hakulinen R, Salkinoja-Salonen M (1982b)Treatment of pulp and paper industry wastewaters in an anaerobic fluidized-bed reactor. Proc Biochem 17(2):18–22

Junna J, Lammi R, Miettinen V (1982) Removal of organic and toxic substances from debarking and Kraft pulp bleaching effluents by activated sludge treatment. Publication of Water Research Institute, National Board of Waters, Finland, No. 49

Jurgensen SJ, Benjamin MM, Ferguson JF (1985) Treatability of thermomechanical pulping process effluents with anaerobic biological reactors. In: Proceedings of Tappi 1985 environmental conference, Tappi Press, Atlanta, GA, USA, pp 83–92

Kaser LE, Dixon DG, Hodsor PV (1984) QSAR studies on chlorophenols, chlorobenzenes and parasubstituted phenols. QSAR in Environmental Toxicology (Kaiser KLE ed.) D. Reidel Publishing Co. Boston, USA, 189–206

Leach JM, Thakore AN (1976) Toxic constituents of mechanical pulping effluents. Tappi J 59 (2):129–132

Lee JW (1993) Anaerobic treatment of pulp and paper mill wastewaters. In: Springer AM (ed) Industrial environmental control, pulp and paper industry. Tappi Press, Atlanta, GA, USA, pp 405–446

Lettinga G, Field JA, Sierra-Alvarez R, vanLier JB, Rintala J (1991) Future perspectives for the anaerobic treatment of forest industry wastewaters. Water Sci Technol 24(3/4):91–102

McDermott GN (1983) Assessing anaerobic treatment potential. In: Symp. anaerobic biotechnology: reducing the cost of industrial wastewater treatment, Argonne Lab. Washington D.C., 14 April 1983

Mikesell MD, Boyd SA (1986) Complete reductive dechlorination and mineralization of pentachlorophenol by anaerobic microorganisms. Appl Environ Microbiol 52:861–865

Qui R, Ferguson JF, Benjamin MM (1988) Sequential anaerobic and aerobic treatment of Kraft pulping wastes. Water Sci Technol 20(1):107–120

Rintala JA, Puhakka JA (1994) Anaerobic treatment in pulp and paper mill. Bioresour Technol 47:1–18

Roger IH (1973) Isolation and chemicals identification of toxic components of Kraft mill wastes. Pulp & Paper Mag Canada 74:111–116

Salkinoja-Salonen M, Valo R, Apajalahati J, Hakulinen R, Silakoski L, Jaakola T (1984) Biodegradation of chlorophenolic compounds in wastes from wood processing industry. In: Klug MJ, Reddy CA (eds) Current perspective in microbial ecology. American Society for Microbiology, Washington, DC, pp 668–676

Schink B (1985) Degradation of unsaturated hydrocarbons by methanogenic enrichment cultures. FEMS Microbiol Ecol 31:69–77

Schink B (1988) Principles and limits of anaerobic degradation: environmental and technological aspects. In: Zehnder AJB (ed) Biology of anaerobic microorganisms. Wiley, New York, pp 771–846

Sierra-Alvarez R, Kortekaas S, van Ekert M, Lettinga G (1991) The anaerobic biodegradability and methanogenic toxicity of pulping wasteswaters. Water Sci Technol 24(3):113–125

Speece RE (1983) Anaerobic biotechnology for industrial wastewater treatment. Environ Sci Technol 17(9):416–426

Temmink JHM, Field JA, van Haastrecht JC, Merckelbach RCM (1989) Acute and sub-acute toxicity of bark tannins to carp (*Cyprinus carpio* L). Water Res 23:341–344

Tiedje JM, Boyd SA, Fathepure BZ (1987) Anaerobic degradation of chlorinated aromatic hydrocarbons. Dev Ind Microbiol 27:117–127

van Lier JB, Mahmoud N, Zeeman G (2008) Anaerobic biological wastewater treatment. In: Henze M, van Loosdrecht MCM, Ekama GA, Brdjamovic D (eds) Biological wastewater treatment: principles, modeling and design. IWA Publishing, London

Welander T, Anderson PE (1985) Anaerobic treatment of wastewater from the production of chemithermomechanical pulp. Water Sci Technol 17(1):103–112

Wilson RW, Murphy KL, Frenette EG (1987) Aerobic and anaerobic treatment of NSSC and CTMP effluent. Pulp & Paper Canada 88(1):T4–8

Wood SL, Ferguson JF, Benjamin MM (1989) Characterization of chlorophenol and chloromethoxybenzene biodegradation during anaerobic treatment. Environ Sci Technol 23:62–68

Zeikus JG, Wellstein AL, Kirk TK (1982) Molecular basis for the biodegradative recalcitrance of lignin in anaerobic environments. FEMS Microbiol Lett 15:193–197

Chapter 5
Anaerobic Reactors Used for Waste Water Treatment

Abstract Different types of reactor configurations used for the anaerobic treatment of wastewaters are presented in this chapter. Anaerobic lagoon, anaerobic contact process, anaerobic filter, upflow anaerobic sludge blanket reactor, fluidized bed reactor, expanded granular sludge bed reactor, internal circulation reactor, anaerobic baffled reactor, membrane coupled high-rate and CSTR systems, anaerobic membrane bioreactors are being used.

Keywords Anaerobic reactors · Waste water treatment · Reactor configurations · Anaerobic lagoon · Anaerobic contact process · Anaerobic filter · Upflow anaerobic sludge blanket reactor · Fluidized bed reactor · Expanded granular sludge bed reactor · Internal circulation reactor · Anaerobic baffled reactor · Membrane coupled high-rate and CSTR system · Anaerobic membrane bioreactors

Many different types of reactor configurations, have been used for the anaerobic treatment of wastewaters (Allen and Liu 1998; Lee 1993; Speece 1983; Kosaric and Blaszczyk 1992; Lee et al. 1989). These are—anaerobic lagoon, anaerobic contact process, anaerobic filter, upflow anaerobic sludge blanket reactor, fluidized bed reactor, expanded granular sludge bed reactor, internal circulation reactor, anaerobic baffled reactor, membrane coupled high-rate and CSTR systems (anaerobic membrane bioreactors). Several variations in the basic designs have been proposed in the literature of which few made it to commercial scale application (McCarty 2001; Van Lier et al. 2015). Presently, the high-rate sludge bed reactors, i.e. UASB and EGSB reactors and their derivatives, are most widely used for the anaerobic treatment of industrial wastewater, having about 90% of the market share of all installed systems (van Lier et al. 2008). Their popularity for treating industrial wastewaters can be attributed to their ease of operation, compactness while using high VLRs at low HRTs (Rajeshwari et al. 2000; van Lier et al. 2008). Membrane coupled high-rate anaerobic reactor configurations are being studied in the recent years, because of the large amount of comparable knowledge from aerobic MBR operations and the application niche which exists for these systems (Dereli et al. 2012). Membrane assisted sludge retention ensures the accumulation of very slowly

© The Author(s) 2017
P. Bajpai, *Anaerobic Technology in Pulp and Paper Industry*, SpringerBriefs in Applied Sciences and Technology, DOI 10.1007/978-981-10-4130-3_5

growing micro-organisms with inferior adherence properties, that are frequently required for the anaerobic treatment of toxic and recalcitrant wastewaters. Recently, van Lier et al. (2015), discussed the evolution of anaerobic sludge bed technology for the treatment of industrial wastewaters in the last forty years, focusing on granular sludge bed systems.

5.1 Anaerobic Lagoon or Covered Lagoon Reactor

Anaerobic lagoons are basically large unsophisticated, low-rate anaerobic reactors (Fig. 5.1). Anaerobic lagoon was first used in the food processing industry in Australia in the 1940s (Springer 1993; Simon and Ullman 1987; Lee 1993) and is the oldest low-rate anaerobic treatment process. It is not widely used in the pulp and paper industry. An anaerobic lagoon is a deep impoundment, essentially free of dissolved oxygen, which promotes anaerobic conditions. The process typically takes place in deep earthen basins, and such ponds are used as anaerobic pre-treatment systems. Anaerobic lagoons are not aerated, heated, or mixed (EPA 2006; Hamilton 2012; Saele 2004). The typical depth of an aerated lagoon is higher than eight feet, with higher depths preferred. Such depths reduce the effects of oxygen diffusion from the surface and allows anaerobic conditions to predominate. In this respect, anaerobic lagoons are different from shallower aerobic or facultative lagoons, making the process similar to that experienced with a single stage unheated anaerobic digester, except that anaerobic lagoons are in an open earthen basin. Furthermore, conventional digesters are typically used for sludge stabilization in a

Fig. 5.1 Anaerobic lagoons. Based on mebig.marmara. edu.tr/Enve424/Chapter7.pdf

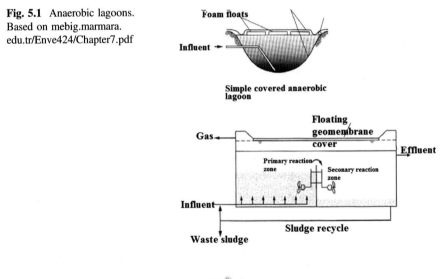

ADI-BV® lagoon process

treatment process, whereas lagoons typically are used to pretreat raw wastewater. The operating cost is lower compared to other alternatives. It is also suitable for wastewaters that contain high levels of suspended solids or significant amounts of oil and grease. The accumulation of a settled biomass sludge results in a very long effective SRTs and maximizes the endogenous destruction of particulate to reduce the amount of sludge requiring disposal. Nutrients released from endogenous decay of the sludge become available for reuse by the active microorganisms. If the anaerobic treatment stage is followed by an aerobic treatment system, waste aerobic sludge can be returned to the covered lagoon for anaerobic digestion. Thus, the total quantity of biosludge requiring disposal from a compared anaerobic/aerobic treatment system is reduced. Periodically accumulated sludge may need to be removed from the process for final disposal. The in-ground construction and the insulated membrane cover allows long hydraulic retention times to be used in a covered lagoon system without greatly reducing process efficiencies due to heat loss. Typical HRTs in a covered lagoon may be between six and thirty days. Corresponding organic loading rates are usually less than 1 or 2 kg $COD/m^3/day$. The low-rate nature of the covered lagoon renders sludge settleability less important than in an anaerobic contact process. The large reactor volumes provide a good degree of equalization for toxics and organic shock loads. However, the process may suffer from mixing inefficiencies and non ideal contact between incoming wastewater and the anaerobic biomass. For many pulp and paper mill applications, minimum hydraulic retention times of 7–10 days would be required for achieving BOD_5 reductions in the range of 75–90%. Solid removal from the lagoon may be required at some time, depending upon the quantity of inorganic solids and the degradability of the suspended material in the influent. Table 5.1 shows the advantages and disadvantages of anaerobic lagoons.

5.2   Anaerobic Contact Reactor

The anaerobic contact process (ACP) was developed in 1950s and was first high rate anaerobic treatment system (Lee 1993). The first anaerobic contact process was reported for the treatment of dilute packing house waste having a COD of about 1300 mg/l (Schroepfer et al. 1955). The various versions of the first generation of these high-rate anaerobic contact process (ACP) systems for medium strength wastewaters were not much successful. The main difficulty was a poor separation of the sludge from the treated water in the secondary clarifier. Other problems were biogas formation and attachment in the settling tank (Rittmann and McCarty 2001). The poor sludge separation was attributed to the very rigorous agitation applied in the bioreactor, creating very small sludge particles having a poor settleability. In addition, super-saturation of solubilized gases resulted in buoyant upward forces in the clarifier. The idea of the very intensified mixing was to ensure optimized contact between the sludge and the wastewater. In the recent years, the ACP systems which have been developed use milder mixing conditions, whereas degasifying units are

Table 5.1 Advantages and disadvantages of anaerobic lagoons

Advantages
More effective for rapid stabilization of strong organic wastes, making higher influent organic loading possible
Produce methane, which can be used to heat buildings, run engines, or produce electricity, but methane collection increases operational problems
Produce less biomass per unit of organic material processed. This equates to savings in sludge handling and disposal costs
Do not require additional energy, because they are not aerated, heated, or mixed. Less expensive to construct and operate. Ponds can be operated in series
Disadvantages
Relatively large area of land is required
Undesirable odors are produced unless provisions are made to oxidize the escaping gases
Gas production should be minimized (sulfate concentration must be reduced to less than 100 mg/L) or mechanical aeration at the surface of the pond to oxidize the escaping gases is necessary
Aerators must be located for ensuring that anaerobic activity is not inhibited by introducing dissolved oxygen to depths below the top 0.6–0.9 m of the anaerobic lagoon. Another option is to locate the lagoon in a remote area
Relatively long detention time is required for organic stabilization due to the slow growth rate of the methane formers and sludge digestion
Seepage of wastewater into the groundwater may be a problem. This problem can be prevented by providing a liner for the lagoon
Environmental conditions directly affect operations so any variance limits the ability to control the process

citeseerx.ist.psu.edu/viewdoc/download?doi=10.1.1.461.8025&rep=rep1

often equipped before the secondary clarification. The modern ACP systems are very effective for concentrated wastewaters with relatively high concentrations of suspended solids. According to van Lier et al. (2008), ACP have a consolidated market share within the full scale applied anaerobic high-rate systems. However, ACP effluents need a subsequent treatment step in order to comply with effluent restrictions.

Anaerobic contract reactor is an outgrowth of the anaerobic lagoon and is similar to the activated sludge process. Separation of the sludge from the settling tank is the critical factor for maintaining high biomass concentration and for operating the contact process. It consists of fully mixed anaerobic reactor and sludge settling tank (Fig. 5.2). A portion of the sludge is returned to the contact reactor to maintain high biomass concentration (3000–10,000 mg/l) in the reactor. Due to the recycling of sludge, the SRT can be controlled to be much longer than the HRT. Separation of the sludge from the settling tank is the critical factor for maintaining high biomass concentration and for operating the contact process. This system is suitable for treating effluents containing a high concentration of suspended solids. It can be operated at an organic loading from 1 to 2 kg BOD/m^3/day.

Fig. 5.2 Anaerobic contact process. Based on Agbalakwe (2011); mebig.marmara.edu.tr/Enve424/Chapter7.pdf

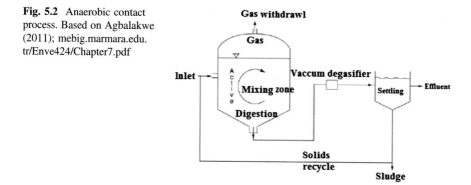

Anaerobic contact process can be applied to a wide range of wastewater concentrations. Although the lower economically practical limit of wastewater concentration is probably in the range of 1000–2000 mg COD/l, there is no well-established upper concentration limit. At very high wastewater concentrations, the completely mixed anaerobic reactor is the best alternative for efficient digestion while reducing internal reactor hydraulic inefficiencies. Wastewaters containing up to 100,000 mg COD/l can be treated in an anaerobic contact process as long as the anaerobic floc produced has satisfactory settling properties. In practice, the floc settleability can be diminished by the presence of high concentrations of dissolved solids. If the untreated wastewater contains significant concentrations of poorly biodegradable suspended solids, then a biomass recycle system can lead to the accumulation of inert solids in the reactor. Over long periods, the accumulation of inert material may cause the displacement of active anaerobic biomass from the process.

The treatment efficiency of an anaerobic contact process is usually much greater than that of a completely mixed digester. Total COD reductions of 90–95% are possible for highly biodegradable wastewaters with COD concentrations of 2–10 g/l. Typical organic loading rates in anaerobic contact systems are between 0.5 and 10 kg COD/m^3/day with HRT of 0.5–5 days.

5.3 Upflow Anaerobic Sludge Blanket Reactor

The upflow anaerobic sludge blanket (UASB) reactor was developed during the 1970s by Lettinga et al. (1976, 1979, 1980, 1987) in Netherlands. This is one of the most remarkable and significant developments in high-rate anaerobic treatment technology. It is basically a tank with a sludge bed (Fig. 5.3) (Gómez 2011; Lettinga et al. 1979). In this reactor, the mixing between sludge and the feedstock is obtained by an even flow distribution combined with a sufficiently high flow velocity and the agitation resulting from gas formation (Lettinga 1995; Duncan Mara 2003). The development of sludge into high-density granules results in the

Fig. 5.3 UASB reactor.
Based on Agbalakwe (2011)

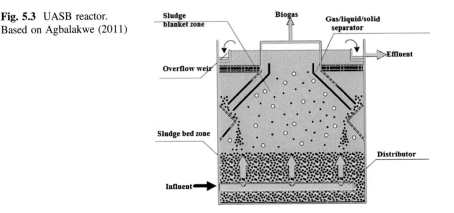

formation a blanket or granular matrix, which is kept in suspension by controlled upflow velocity (Duncan Mara 2003).

The sludge retention in such a reactor is based on the formation of well settleable sludge aggregates (flocs or granules), and on the application of a reverse funnel-shaped internal gas–liquid–solids separation device. Many successful performance results have been reported at laboratory and pilot-scale applications using anaerobic granular sludge bed processes, which resulted in the establishment of thousands of full-scale reactors worldwide (Nnaji 2013; Lim and Kim 2014; van Lier et al. 2008). Undoubtedly, anaerobic sludge bed reactors, are by far the most popular anaerobic wastewater treatment systems so far, having a wide application potential in industrial wastewater treatment. The first UASB reactors were installed for the treatment of food, beverage and agro-based wastewaters, rapidly followed by applications for paper and board mill effluents in 1983 (Habets and Knelissen 1985). Most of the full-scale reactors are used for treating agro-industrial wastewater, but the applications for the treatment of wastewaters from chemical industries are increasing (van Lier et al. 2008; Rajagopal et al. 2013). The wastewater moves in an upward flow through the UASB reactor. Good settle-ability, low HRTs, high biomass concentrations (up to 80 g l^{-1}), effective solids/liquid separation, and operation at high VLRs can be achieved by UASB reactor systems (Speece 1996). The design VLR is typically in the range of 4–15 kg COD m^{-3} day (Rittmann and McCarty 2001). One of the major limitations of this process is related to wastewaters having a high suspended solids content, which hampers the development of dense granular sludge (Alphenaar 1994).

UASB can treat various concentrations ranging from 250 to 24,000 mg/l COD of wastewaters including various pulp mill effluents. The high biomass concentration also renders UASB to be more tolerable to toxicants. Loading rates generally range from 3.5 to 5 kg BOD/m^3/day and can be up to 8 kg BOD/m^3/day. UASB has several advantages compared with other high-rate anaerobic systems (Table 5.2). The capital costs for the UASB reactor are comparatively lower than for other anaerobic systems. A high loading rate means reduced reactor volumes

Table 5.2 Advantages of UASB reactors	Availability of granular or flocculent sludge, thus no requirement of a support medium
	High biomass content, enabling a wide range of loading rates and high COD removal efficiency
	Blanketing of sludge, enabling short hydraulic retention time and high solids retention time
	Rising gas bubbles produced, eliminating the need of mixing and thus lower energy demand
	Long experience in practice
	Weiland and Rozzi (1991), Zoutberg and de Been (1997), Hickey et al. (1991)

Table 5.3 Challenges of UASB reactors	Start-up is susceptible to temperature and organic shock loads
	Difficulties in controlling the bed expansions, thus limiting the applicable organic loading rates
	Wash-out, flotation and disintegration of granular sludge
	Performance deteriorates at low temperatures
	High sulphate concentration
	Necessity of post-treatment to reach the discharge standards for organic matter, nutrients and pathogens
	Purification of biogas
	Weiland and Rozzi (1991), Lettinga and Hulshoff Pol (1991), Li et al. (2008), Lew et al. (2011), Heffernan et al. (2011), Mahmoud et al. (2008)

and the separation of gas, liquid and solid often only needs to take place in one tank. Also, no support medium is required for attachment of the biomass. The UASB reactor has comparatively low energy, chemical and labour requirements. If the reactor is seeded with adapted granular sludge from another full-scale plant treating a similar waste, start-up can be very rapid. The challenges of UASB reactors are presented in Table 5.3.

5.4 Anaerobic Filter Reactor

The anaerobic filter (AF) (Hamilton 2012) are also known as the fixed film digester or packed bed digester. These reactors were already applied in the nineteenth century (McCarty 2001) but the application for industrial wastewater treatment started in the 1960s in the United States (Young and McCarty 1969; Young 1991; Young and Yang 1989). Since 1981, about 130–140 full-scale upflow AF installations have been put in operation for the treatment of various types of wastewater, which is about 6% of the total amount of installed high-rate reactors. The experiences with the system certainly are rather satisfactory; applying modest to relatively high loading rates up to 10 kg COD m^{-3} day^{-1}. AF technology has been widely applied for treatment of wastewaters from the beverage, food-processing, pharmaceutical and chemical industries due to its high capability of biosolids retention (Ersahin et al. 2007). The AF system will remain attractive for treatment of mainly soluble types of wastewaters, particularly when the sludge granulation process cannot occur satisfactory. On the other hand, long-term problems related to system clogging and the stability of filter material caused a decline in the number of installed full-scale AF systems.

This reactor relies upon a media substrate to retain the microorganisms within the reactor vessel, and the filter material is usually made from ceramics, glass, plastic, or wood (EPA 2006). As the growth of microorganisms requires relatively long periods of time to develop, their holding in the reactor by the media can facilitate the anaerobic digestion process (Gerardi 2003).

The AF has been developed as a biofilm system in which biomass is retained based on the attachment of a biofilm to the stationary carrier material; entrapment of sludge particles between the interstices of the packing material, and the sedimentation and formation of very well settling sludge aggregates (Fig. 5.4). AF technology can be applied in upflow and downflow reactors (Young and Yang 1989). Various types of synthetic packing materials, as well as natural packing materials, have been investigated in order to be used in AFs. These are gravel, coke and bamboo segments. Important aspects of the packing materials are shape, size, weight, specific surface area, and porosity. Also the surface adherence properties with regard to bacterial attachment are important. Applying proper support material, AF systems can be rapidly started, because of the efficient adherence of anaerobic organisms to the inert carrier. The ease of starting up the AFs was the main reason for its popularity in the eighties and nineties. Problems with AF systems generally occur during long-term operation. The major disadvantage of the AF concept is the

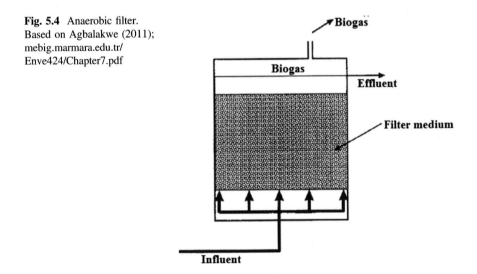

Fig. 5.4 Anaerobic filter. Based on Agbalakwe (2011); mebig.marmara.edu.tr/ Enve424/Chapter7.pdf

difficulty to maintain the required contact between sludge and wastewater, because clogging of the "bed" easily occurs. This is particularly the case for partly soluble wastewaters. These clogging problems can be partly overcome by applying a primary settler and/or a pre-acidification step (Seyfried 1988). However, this would require the construction and operation of additional units. Moreover, apart from the higher costs, it would not completely eliminate the problem of short-circuiting (clogging of the bed), leading to disappointing treatment efficiencies.

5.5   Anaerobic Fluidized and Expanded Bed Systems

These reactors are regarded as the second generation of anaerobic sludge bed reactors which achieve very high VLRs. In the lab scale, 30–60 kg COD m^{-3} day^{-1} and at full scale: 20–40 kg COD m^{-3} day^{-1} have been obtained. The fluidized bed (FB) system can be regarded as an advanced anaerobic technology which may reach loading rates exceeding 40 kg COD m^{-3} day^{-1}, when operated under defined conditions (Moletta et al. 1994; Heijnen et al. 1990; Li and Sutton 1981). The FB process is based on the occurrence of bacterial attachment to non-fixed or mobile carrier particles, which consist, of fine sand, basalt, pumice, or plastic (Fig. 5.5). FB reactors are very efficient due to following reasons:

– Good mass transfer resulting from liquid turbulence and high flow rate around the particles
– Less short circuiting and less clogging due to the occurrence of large pores through bed expansion
– High specific surface area of the carriers make FB reactors highly efficient

Fig. 5.5 Fluidized bed reactor. Based on Agbalakwe (2011); mebig.marmara.edu.tr/Enve424/Chapter7.pdf

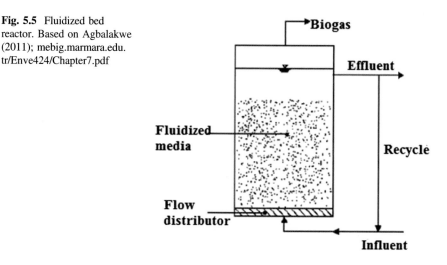

Long-term stable operation however, appears to be problematic. The system depends on the formation of a more or less uniform attached biofilm and/or particles. Ehlinger (1994) has reported that to maintain a stable situation with respect to the biofilm development, pre-acidification is important and dispersed matter should be absent in the feed. Inspite of that, an even film thickness is very difficult to control and in many cases segregation of different types of biofilms over the height of the reactor occurs. In case of full-scale reactors, bare carrier particles may separate from the biofilms leading to operational problems. In order to sustain the biofilm particles in the reactor, adjustments of the flow are required, after which the support material will start to collect in the lower part of the reactor as a kind of stationary bed, while in contrast light fluffy aggregates will be present in the upper part of the reactor. Retention of these fluffy aggregates can only be performed when the superficial velocity remains relatively low, which is not the aim of an FB system.

Modern FB reactors like the Anaflux system depend on bed expansion instead of bed fluidization (Holst et al. 1997). The bed expansion allows a much wider distribution of prevailing biofilms therefore, the system is easy to operate. An inert porous carrier material is used for bacterial attachment in the Anaflux system. The reactor uses a triple phase separator at top of the reactor which is almost similar to the Gas liquid solids separator device in UASB reactors. When the biofilm layer attached to the media becomes excessively over-developed and the concerning aggregates subsequently collect in the separator device, the material is extracted from the reactor periodically by an external pump, in which it is subjected to enough shear to remove part of the biofilm. Then, both the media and detached biomass are returned to the reactor; the free biomass is then allowed to get washed out from the system. The density of the media is controlled in this way and a more homogeneous reactor bed is created. Up to 30–90 kg volatile suspended solids m^{-3}, reactor can be retained in this way and because of the applied high liquid upflow velocities, i.e. up to 10 m h^{-1}, an excellent liquid-biomass contact is achieved. The system can be applied to wastewaters with a suspended solids concentration of <500 mg/l. Most of the full-scale anaerobic FB reactors are installed as Anaflux processes. Nevertheless, at present, the EGSB reactors are much more of commercial interest for full scale applications than the more expensive FB systems (Driessen and Vereijken 2003). EGSB reactors can be defined as a modification of the UASB reactor in which the granules are partially fluidized by effluent recycle at a liquid upflow velocity of 5–6 m/h (Frankin and Zoutberg 1996). This reactor shows better mass transfer characteristics over the UASB reactor (Fig. 5.6) (Mutombo 2004).

A special version of the EGSB concept is the Internal Circulation reactor (IC) (Vellinga et al. 1986) (Fig. 5.7). The biogas produced is separated from the liquid halfway the reactor by gas/liquid separator device and conveyed upwards through a pipe to a degasifier unit. The separated gas is removed from the reactor and the sludge-liquid mixture drops back to the bottom of the reactor through a different pipe. This gas lift transport results to an improved contact between the sludge and wastewater (Vellinga et al. 1986; Pereboom and Vereijken 1994; Habets et al. 1997). The IC reactor can be considered as two anaerobic treatment compartments (like UASB) on top of each other, one highly loaded and the other with

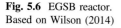

Fig. 5.6 EGSB reactor.
Based on Wilson (2014)

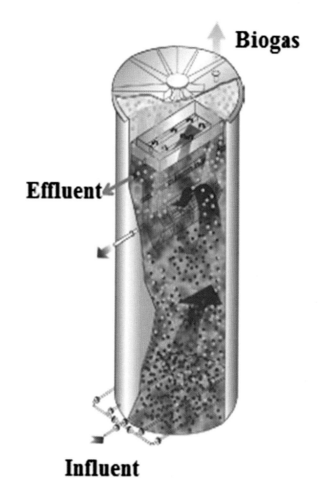

low loading (Mutombo 2004). A special feature associated with the IC reactor is related to its highly efficient multi-level circulation system. The IC technology is based on the proven UASB process (Habets 2005). Typically, the loading rate of the IC reactor can be higher as compared to that of the UASB reactor (Driessen and Vereijken 2003).

5.6 Anaerobic Baffled Reactor

Anaerobic baffled reactor (ABR) is a high rate bioreactor (Fig. 5.8). It was developed by McCarty and co-workers at Stanford University. It is described as a series of upflow anaerobic sludge blanket reactors because it is divided into several compartments (Bachmann et al. 1985; Barber and Stuckey 1999; Zhu et al. 2015).

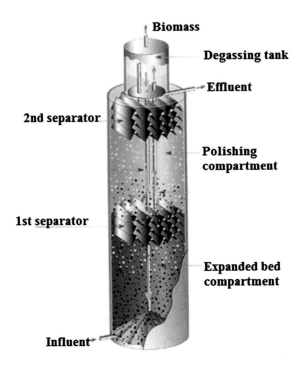

Fig. 5.7 IC reactor. Based on Wilson (2014)

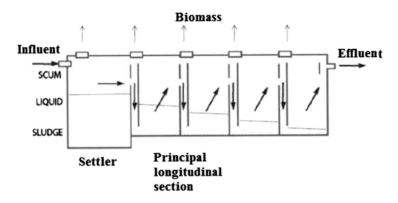

Fig. 5.8 Anaerobic baffled reactor. Based on Agbalakwe (2011)

A typical ABR consists of a series of vertical baffles which direct the wastewater under and over the baffles as it passes from the inlet to the outlet. The washout of bacteria is reduced. This enables the ABR to retain active biological mass without the use of any fixed media. The bacteria within the reactor rise and settle with gas production in each compartment, but they move down the reactor horizontally at a relatively slow rate, giving rise to a SRT of 100 days at a HRT of 20 h. The slow movement in horizontal direction allows wastewater to come into intimate contact with the active biomass as it passes through the ABR with short HRTs of 6–20 h (Bachmann et al. 1985; Barber and Stuckey 1999; Zhu et al. 2015). ABR has a simple design and does not require special gas or sludge separation equipment. This reactor can be used for almost all soluble organic wastewater from low to high strength wastewaters. It could be considered a potential reactor system for treating municipal wastewater in tropical and sub-tropical areas of developing countries considering its simple structure and operation.

5.7 Anaerobic Membrane Reactor

In the recent years a lot of research is being conducted on anaerobic membrane bioreactors (Fig. 5.9). Membrane technology is an interesting option in those areas where established technologies may not succeed. In Anaerobic Membrane Reactor (AMR), the size of reactor is reduced and organic loadings are increased due to higher biomass concentrations. Almost complete capturing of solids (much longer SRT) occurs which result in maximum removal of volatile fatty acids and degradable soluble organics resulting in better quality effluent. The greatest challenge in AMR is the organic fouling. This is typically caused by accumulation of colloidal materials and bacteria on the membrane surface. High liquid velocities across the membrane and gas agitation systems can be used to reduce membrane fouling. High pumping flow rates across the membrane may result in the loss of bacteria due to cell lysis. Developments in membrane design in the recent years and fouling control measures could make AMR a viable technology in future. Currently, only a few full scale AMR systems are in operation. But an increase in this emerging technology is expected considering the sharp drop in membrane prices (Henze 2008; Calli 2010).

Fig. 5.9 Anaerobic membrane bioreactor. Based on Agbalakwe (2011); mebig. marmara.edu.tr/Enve424/ Chapter7.pdf

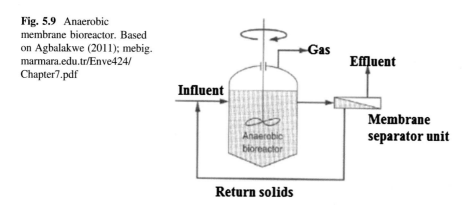

5.8 Hybrid Upflow Anaerobic Sludge Blanket/Anaerobic Filter

Hybrid systems are designed to take advantage of special features of two or more process concepts (Fig. 5.10). Several hybrid reactor configurations which combine UASB/fixed media systems have been developed and evaluated in pilot plants or at full-scale (Lee 1993; Lee et al. 1989). As an example, where adapted granular sludge is not available, the UASB/fixed film hybrid may offer a faster startup than UASB alone. Development and entrapment of a flocculant anaerobic biomass, as well as growth of a fixed biofilm, normally proceed more rapidly than development and growth of granular sludge from an initial flocculant seed. Other features of the UASB/fixed film hybrid anaerobic reactor are—High overall reactor biomass concentrations than UASB alone, resulting in a small reactor volume; greater resistance to toxicity shock loads by having both a granular and a fixed film biomass; where biomass support media cost is high, the combination of processes may offer a capital cost advantage over an anaerobic filter alone, sized to achieve a similar treatment efficiency. The primary disadvantage may be eventual plugging of the fixed media operating in an upflow mode and the potential difficulty of optimizing two processes physically housed in a single vessel for a wide range of flow and loading conditions. Separation of the acid-forming and methane-forming phases into two stages, at least in theory, allows the design and operation of each phase to be optimized independently of each other. The facultative acid-forming bacteria in the first stage can provide significant protection to the more sensitive methanogenic strict anaerobes. This is particularly the case when the oxidants such as hydrogen peroxide are present in the wastewater.

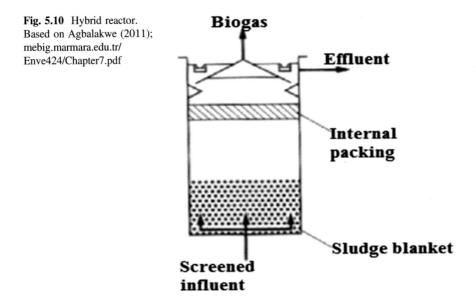

Fig. 5.10 Hybrid reactor. Based on Agbalakwe (2011); mebig.marmara.edu.tr/ Enve424/Chapter7.pdf

References

Agbalakwe E (2011) Anaerobic treatment of glycol contaminated wastewater for methane production. MS thesis. University of Stavanger, Norway

Allen DG, Liu HW (1998) Pulp mill effluent remediation. In: Meyers RA (ed) Encyclopedia of environmental analysis and remediation, vol 6. Wiley, Wiley Interscience Publication, New York, 1998, pp 3871–3887

Alphenaar PA (1994) Anaerobic granular sludge: characterization and factors affecting its functioning. PhD thesis, G. Lettinga (promotor), Department of Environmental Technology, Agricultural University, Wageningen, The Netherlands

Bachmann A, Beard VL, McCarty PL (1985) Performance characteristics of the anaerobic baffled reactor. Water Res 19:99–106

Barber WP, Stuckey DC (1999) The use of the anaerobic baffled reactor (ABR) for wastewater treatment: a review. Water Res 33:1559–1578

Calli B (2010) Lecture notes on ENVE 424 anaerobic treatment. Department of Environmental Engineering, Marmara University Turkey.

Dereli RK, Ersahin ME, Ozgun H, Ozturk I, Jeison D, van der Zee F, van Lier JB (2012) Potentials of anaerobic membrane bioreactors to overcome treatment limitations induced by industrial wastewaters. Bioresour Technol 122:160–170

Driessen W, Vereijken T (2003) Recent development in biological treatment of brewery effluent. In: Proceedings of the institute and guild of brewery convention, Zambia, pp 268–376

Duncan Mara NH (2003) Handbook of water and wastewater microbiology. Academic Press, New York

Ehlinger F (1994) Anaerobic biological fluidized beds: operating experiences in France. In: 7th international symposium on anaerobic digestion, Cape Town, South Africa, 23–27 January

EPA (United States Environmental Protection Agency) (2006) Biosolids technology fact sheet: multi-stage anaerobic digestion. Retrieved 11 Oct 2013. URL: http://water.epa.gov/scitech/wastetech/upload/2006_10_16_mtb_multi-stage.pdf

Ersahin ME, Dereli RK, Insel G, Ozturk I, Kinaci C (2007) Model based evaluation for the anaerobic treatment of corn processing wastewaters. Clean-Soil Air Water 35(6):576–581

Frankin R, Zoutberg GR (1996) Anaerobic treatment of chemical and brewery wastewater with a new type of anaerobic reactor: the biobed EGSB reactor. Water Sci Technol 34(5–6):375–381. doi:10.1016/0273-1223(96)00668-3

Gerardi MH (2003) The microbiology of anaerobic digesters. Wiley, Hoboken, NJ

Gómez RR (2011) Upflow anaerobic sludge blanket reactor modelling. Royal Institute of Technology, Stockholm, Sweden

Habets LHA (2005) Introduction of the IC reactor in the paper industry. http://www.environmentalexpert.com/Files/587/articles/5523/paques16.pdf

Habets LHA, Knelissen JH (1985) Application of the UASB reactor for anaerobic treatment of paper and board mill effluent. Water Sci Technol 17(1):61–75

Habets LHA, Engelaar AJHH, Groeneveld N (1997) Anaerobic treatment of inuline effluent in an internal circulation reactor. Water Sci Technol 35(10):189–197

Hamilton DW (2012) Types of anaerobic digesters. Retrieved 20 Oct 2013. URL: http://www.extension.org/pages/30307/types-of-anaerobicdigesters#. Umo79KkZ3Cz

Heffernan B, van Lier JB, van der Lubbe J (2011) Performance review of large scale up-flow anaerobic sludge blanket sewage treatment plants. Water Sci Technol 63(1):100–107

Heijnen SJ, Mulder A, Weltevrede R, Hols PH, van Leeuwen HLJM (1990) Large-scale anaerobic/aerobic treatment of complex industrial wastewater using immobilized biomass in fluidized bed and air-lift suspension reactors. Chem Eng Technol 13(1):202–208

Henze M (2008) Biological wastewater treatment: principles, modelling and design. IWA Pub, London, p 511

Hickey RF, Wu W-M, Veiga MC, Jones R (1991) Start-up, operation, monitoring and control of high-rate anaerobic treatment systems. Water Sci Technol 24(8):207–255

Holst TC, Truc A, Pujol R (1997) Anaerobic fluidised beds: ten years of industrial experience. Water Sci Technol 36(6–7):415–422

Kosaric N, BlaszczyK R (1992) Industrial effluent processing. In: Lederberg J (ed) Encyclopedia of microbiology, vol 2. Academic Press Inc., New York, pp 473–491

Lee JW (1993) Anaerobic treatment of pulp and paper mill wastewaters. In: Springer AM (ed) Industrial environmental control, pulp and paper industry. Tappi Press, Atlanta, GA, USA, pp 405–446

Lee JM, Peterson DL, Stickney AR (1989) Anaerobic treatment of pulp mill wastewaters. Environ Prog 8(2):73–86

Lettinga G (1995) Anaerobic digestion and wastewater treatment systems. Antonie van Leeuwenhoek 67(1):3–28. doi:10.1007/BF00872193

Lettinga G, Hulshoff Pol LW (1991) UASB-process design for various types of wastewaters. Water Sci Technol 24(8):87e107

Lettinga G, van der Ben J, van der Sar J (1976) Anaerobe zuivering van het afvalwater van de bietsuikerindustrie. H₂O 9:38–43

Lettinga G, van Velsen L, de Zeeuw W, Hobma SW (1979) The application of anaerobic digestion to industrial pollution treatment. In: 1st international symposium on anaerobic digestion, Cardiff, UK, 17–21 September

Lettinga G, van Velsen AFM, Hobma SW, de Zeeuw W, Klapwijk A (1980) Use of the upflow sludge blanket (USB) reactor concept for biological wastewater treatment, especially for anaerobic treatment. Biotechnol Bioeng 22(4):699–734

Lettinga G, Zehnder AJB, Grotenhuis JTC, Hulshoff Pol LW (eds) (1987) In: GASMAT: international workshop on granular anaerobic sludge, microbiology and technology, Lunteren, The Netherlands, PUDOC, Wageningen, The Netherlands, 25–27 October 1987

Lew B, Lustig I, Beliavski M, Tarre S, Green M (2011) An integrated UASB-sludge digester system for raw domestic wastewater treatment in temperate climates. Bioresour Technol 102 (7):4921–4924

Li A, Sutton PM (1981) Dorr oliver anitron system, fluidized bed technology for methane production from dairy wastes. In: Whey products institute annual meeting, Chicago, USA

Li J, Hu B, Zheng P, Qaisar M, Mei L (2008) Filamentous granular sludge bulking in a laboratory scale UASB reactor. Bioresour Technol 99(9):3431–3438

Lim SJ, Kim TH (2014) Applicability and trends of anaerobic granular sludge treatment processes. Biomass Bioenergy 60:189–202

Mahmoud N, Zeeman G, van Lier J (2008) Adapting UASB technology for sewage treatment in Palestine and Jordan. Water Sci Technol 57(3):361–366

McCarty PL (2001) The development of anaerobic treatment and its future. Water Sci Technol 44 (8):149–156. mebig.marmara.edu.tr/Enve424/Chapter7.pdf

Moletta R, Escoffier Y, Ehlinger F, Coudert J-P, Leyris J-P (1994) On-line automatic control system for monitoring an anaerobic fluidized-bed reactor: response to organic overload. Water Sci Technol 30(12):11–20

Mutombo DT (2004) Internal circulation reactor: pushing the limits of anaerobic industrial effluents treatment technologies. In: Proceedings of the 2004 Water Institute of Southern Africa (WISA) biennial conference, Cape Town, South Africa.

Nnaji CC (2013) A review of the upflow anaerobic sludge blanket reactor. Desalin Water Treat 52:4122–4143

Pereboom JHF, Vereijken TLFM (1994) Methanogenic granule development in full scale internal circulation reactors. Water Sci Technol 30(8):9–21

Rajagopal R, Saady NMC, Torrijos M, Thanikal JV, Hung YT (2013) Sustainable agro-food industrial wastewater treatment using high rate anaerobic process. Water 5:292–311

Rajeshwari KV, Balakrishnan M, Kansal A, Lata K, Kishore VVN (2000) State-of-the-art of anaerobic digestion technology for industrial wastewater treatment. Renew Sustain Energy Rev 4:135–156

Rittmann BE, McCarty PL (2001) Environmental biotechnology: principles and applications. McGraw-Hill, New York

Saele LM (2004) Covered lagoons, AgSTAR national conference. Retrieved 22 Oct 2013. URL: http://www.epa.gov/agstar/documents/conf04/saele.pdf

Schroepfer GJ, Fullen WJ, Johnson AS, Ziemke NR, Anderson JJ (1955) The anaerobic contact process as applied to packing house wastes. Sew Ind Wastes 27(4):460–486

Seyfried CF (1988) Reprints verfahrenstechnik abwasserreiningung, GVC-Diskussionstagung, Baden-Baden, Germany, 17–19 Oct

Simon O, Ullman P (1987) Present state of anaerobic treatment. Paperi Ja Puu–Paper Och Tra 1987(6):510–515

Speece RE (1983) Anaerobic biotechnology for industrial wastewater treatment. Environ Sci Technol 17(9):416–426

Speece RE (1996) Anaerobic biotechnology for industrial wastewaters. Archae Press, USA

Springer AM (1993) Bioprocessing of pulp and paper mill effluents-past, present and future. Paperi Ja Puu–Paper Timber 75(3):156–161

van Lier JB, Mahmoud N, Zeeman G (2008) Anaerobic biological wastewater treatment. In: Henze M, van Loosdrecht MCM, Ekama GA, Brdjamovic D (eds) Biological wastewater treatment: principles, modeling and design. IWA Publishing, London

van Lier J, Van der Zee F, Frijters C, Ersahin M (2015) Celebrating 40 years anaerobic sludge bed reactors for industrial wastewater treatment. Rev Environ Sci Bio/Techno 14:681–702

Vellinga SHJ, Hack PJFM, van der Vlugt AJ (1986) New type "high rate" anaerobic reactor; first experience on semitechnical scale with a revolutionary and high loaded anaerobic system. In: Anaerobic treatment: a grown-up technology, aquatech water treatment conference, Amsterdam, The Netherlands, 15–19 Sept

Weiland P, Rozzi A (1991) The start-up, operation and monitoring of high-rate anaerobic treatment systems: discusser's report. Water Sci Technol 24(8):257–277

Wilson DR (2014) www.seai.ie/...Energy.../Waste-to-Energy—Anaerobic-digestion-for-large-industry.p

Young JC (1991) Factors affecting the design and performance of upflow anaerobic filters. Water Sci Technol 24(8):133–155

Young JC, McCarty PL (1969) The anaerobic filter for waste treatment. J Water Pollut Control Fed 41:160–173

Young JC, Yang BS (1989) Design considerations for full-scale anaerobic filters. J Water Pollut Control Fed 61(9):1576–1587

Zhu G, Zou R, Jha AK, Huang X, Liu L, Liu C (2015) Recent developments and future perspectives of anaerobic baffled bioreactor for wastewater treatment and energy recovery. Crit Rev Environ Sci Technol 45(12):1243–1276

Zoutberg GR, de Been P (1997) The Biobed_ EGSB (expanded granular sludge bed) system covers shortcomings of the upflow anaerobic sludge blanket reactor in the chemical industry. Water Sci Technol 35(10):183–187

Chapter 6
Pulp and Paper Making Process

Abstract Pulp and paper mills are highly complex and integrate many different process areas including Raw material preparation (wood debarking and chips preparation); Pulping (including cooking or refining, washing, screening and cleaning and thickening); Pulp bleaching (if required) and Paper making. The major unit operations are pulping, bleaching and paper making. Brief description of pulp and paper making process is presented in this chapter.

Keywords Pulp and paper mills · Raw material preparation · Wood debarking · Chips preparation · Pulping · Cooking · Refining · Washing · Screening · Cleaning · Thickening · Pulp bleaching · Paper making

Pulp and Paper industry is a highly water-dependent industry when compared with many other industries. Due to the severe environmental regulations, the industry is responsible for the management of the water resources they use. Such resources are usually obtained from the surface and ground waters and after being used in almost all the main process stages (Fig. 6.1), and also for cleaning the equipment, cooling the machines, etc. form the main part of the liquid reject from the pulp and paper industry (Kamali et al. 2016). Due to the increasing concerns on the scarcity of water resources, the water management in water intensive industry is of high importance and therefore strict environmental regulations have been developed for ensuring the sustainable use of the water resources in industrial water users. At the beginning of the last century the manufacturing processes in addition to other internal use required high amount of water (200–500 m^3/tonne paper). But now, this amount has been significantly reduced due to the technological advances in the pulp and paper production processes. Furthermore, in many developed countries, the use of recovered paper produced has significantly increased in the last two decades resulting in a decrease in the amount of the wastewater generated for the production of pulp and paper, due to the recycled fiber mills being less water intensive as compared to virgin mills (Hong and Li 2012). Although the industry is a large user of water, only a small part of the water used is utilized during the manufacturing activities in a typical pulp and paper mill. Wiegand et al. (2011)

© The Author(s) 2017
P. Bajpai, *Anaerobic Technology in Pulp and Paper Industry*, SpringerBriefs in Applied Sciences and Technology, DOI 10.1007/978-981-10-4130-3_6

have reported that in United States, about 88% of the intake water is returned to the surface waters after being treated, whereas only 11% of it is evaporated and 1% is embedded in products or in solid wastes. Therefore, advanced treatment processes can significantly aid the mills to improve the quality of the effluents satisfying the environmental regulations. Furthermore, some internal treatment processes can be provided in order to re-use the water during the manufacturing processes.

Pulp and paper manufacture is a complex process. The basic unit operations are (Fig. 6.1):

– Raw material preparation (wood debarking and chips preparation).
– Pulping (including cooking or refining, washing, screening and cleaning and thickening).
– Pulp bleaching (if required).
– Paper making.

The three major unit operations are pulping, bleaching and paper making. A large amount of water is used in these processes. A mill may have one, two or all three of these basic operations and often more than one of a given operation. The major sources of water effluents are shown in Table 6.1.

Wood pulp is produced by three main processes: Mechanical forces in the presence of water (mechanical pulping). The process involves passing a block of wood, usually debarked, through a rotating grindstone where the fibres are stripped of and suspended in water; Chemical pulping utilises significantly large amounts of chemicals to break down the wood in the presence of heat and pressure. The spent liquor is then either recycled or disposed of by burning for heat recovery; Chemical thermo-mechanical pulping is the combination of chemical and mechanical pulping. The wood is first partially softened by chemicals and the remainder of the pulping proceeds with mechanical force (Bajpai 2012; Thompson et al. 2001).

Fig. 6.1 Schematic of pulp and paper production process

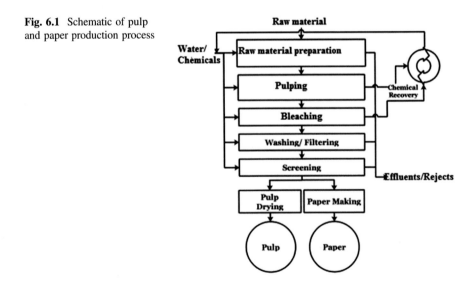

Table 6.1 Major sources of water effluents

Waste water from material preparation
Spent pulping liquor from pulp washing
Digester and evaporator condensates (chemical pulping only)
White waters from screening, cleaning and thickening systems
Bleach plant washers filtrates
Paper machine white water

In the 1970s and 1980s, there was concern over the release of chlorinated organic substances, such as dioxins and furans, from the use of chlorine in pulp bleaching. Facing market and environmental demands for "Elementary Chlorine Free" (ECF) and "Totally Chlorine Free" (TCF) bleached pulps, mills adopted bleaching processes which use chlorine dioxide (ECF pulp) or which use oxygen-containing compounds such as molecular oxygen, peroxide and ozone (TCP pulp) (Lovblad 1999).

In the production of paper, pulp is diluted to at least 99% with water and a mineral filler; china clay, titanium dioxide or chalk; and water-soluble substances such as optical brighteners and polyvinyl alcohol are added (Hentzschel et al. 1998). This is then pumped to a headbox and is distributed evenly along a moving wire cloth. This even distribution is facilitated by the constant side-to-side movement and vibration afforded by the headbox. The majority of the water drains through the wire leading to the formation of a wet paper sheet. This is then vacuum dried and pressed, to extract more water and form the paper sheet. Residual water is removed by passing it through a series of steam-heated cylinders.

Recycled paper is an important source of cellulose fibre for certain paper and board grades (corrugated paper, newsprint). For white grades, such as newsprint, the recycled fibre is de-inked using flotation, followed by washing and screening. Soluble components such as starch are removed in the wastewater.

In case of an integrated pulp and paper mill, most of white water from paper making operation is recycled to the pulp stock preparation and pulp washing in the pulping operation; thus the mill effluent is mainly from the spent pulping liquor, evaporator condensates and bleaching washer filtrates, although a small amount of effluent may be produced from some other operations, such as raw material preparation. Those mills which do not perform pulping and bleaching, the white water from papermaking is the only source of the mill effluent. The pollution loads from the pulp and paper industry mainly depends on pulping and bleaching methods used (Bajpai 2012).

For each tonne of manufactured pulp, the waste water discharge volume ranges from 30 to 180 m^3 whereas 20–70 m^3 is discharged per tonne of paper and paperboard (Gullichsen 1991; Miner and Unwin 1991). The quantities and characteristics of the generated pulp and paper waste waters are highly dependent on the type of raw material, the process conditions applied such as temperature, pH, pressure, chemical and mechanical treatments and the specific water consumption (Stemberg and Norberg 1977). Chemical addition and, to a lesser extent, high temperatures and pressure result in an increased release of organic matter into the process water and extensive solubilization of lignin. Therefore, the pollution loads and the colour due to dissolved lignin compounds is very high for chemical as compared to mechanical pulping effluents (Corson and Lloyd 1978; Virkola and Honkanen 1985). The COD loads associated with the mechanical pulping processes range from 20 to 50 kg COD/tonne of pulp whereas in case of soda pulping processes the COD loads may be as high as 500–900 kg COD/tonne of pulp (Stemberg and Norberg 1977; Anonymous 1986). Nevertheless, the black liquors originating from Kraft and soda processes are usually burnt to recover the pulping chemicals and the calorific power from the organic components. This reduces to a great extent, the environmental impact associated with these pulping processes. Conventional recovery processes are not economically viable in small paper mills and in those mills using nonwoody raw materials with a high silica content. Black liquors represent a very important pollution source in several countries where small scale mills are common (Anonymous 1986; Velasco et al. 1985; Gonenc et al. 1990).

Pulp and paper industry waste waters may cause considerable damage to receiving waters if discharged untreated. The environmental impact associated with these wastewaters is not only restricted to the oxygen demand but also numerous effluents from the pulp and paper industry show acute or chronic toxicity to fish and other aquatic organisms (Roger 1973; Leach and Thakore 1976). Furthermore, these wastewaters streams often exert inhibitory effects on microorganisms, which can disturb biological treatment systems (Bajpai and Bajpai 1997; Bajpai 2013; Ferguson and Benjamin 1985; Welander 1988).

References

Anonymous (1986) Comprehensive industry document for small pulp and paper industry. In: Chakrabarti SP, Kumar A (eds) Comprehensive Industry Document Series COINDS/22/1986. Central Board for the Prevention and Control of Water Pollution, New Delhi, India

Bajpai P (2012) Biotechnology in pulp and paper processing. Springer-Verlag Inc., New York, NY

Bajpai P (2013) Bleach plant effluents from the pulp and paper industry. Springer Briefs in Applied Science and Technology, Springer International Publishing

Bajpai P, Bajpai PK (1997) Reduction of organochlorine compounds in bleach plant effluents. Adv Biochem Eng/Biotechnol (Eriksson KEL ed.) Springer-Verlag 57:213–259

Corson JA, Lloyd JA (1978) Refiner pulp mill effluent. Part I. Generation of suspended and dissolved solid fractions. Pap Puu 60(6–7):407–410

Ferguson JF, Benjamin MM (1985) Studies of anaerobic treatment of sulphite process wastes. Water Sci Technol 17(1):113–122

Gonenc IE, Ilhan R, Orhan D (1990) A rational approach to the design of a rotating disc system for the treatment of black liquor. Water Sci Technol 22(9):207–214

Gullichsen J (1991) Process internal measures to reduce pulp mill pollution load. Water Sci Technol 24:45–53

Hentzschel P, Martin G, Pelzer R, Winkler K (1998) Steps towards optimized paper brightness. Wochenbl Papierfab 126:176–180

Hong J, Li X (2012) Environmental assessment of recycled printing and writing paper: a case study in China. Waste Manage 32(2):264–270

Kamali MR, Gameiro T, Costa MEV, Capela I (2016) Anaerobic digestion of pulp and paper mill wastes—an overview of the developments and improvement opportunities. Chem Eng J 298:162

Leach JM, Thakore AN (1976) Toxic constituents of mechanical pulping effluents. Tappi J 59 (2):129–132

Lovblad R (1999) Clean technology development in kraft pulping—Sodra Cell's strategy for minimising environmental impacts of effluent discharges. Environ Prot Bull Issue 62:3–6

Miner R, Unwin J (1991) Progress in reducing water use and wastewater loads in the US paper industry. Tappi J 74:127–131

Roger IH (1973) Isolation and chemicals identification of toxic components of Kraft mill wastes. Pulp Paper Mag Canada 74:111–116

Stemberg E, Norberg G (1977) Effluent from manufacturing of thermomechanical pulps and their treatment. International Mechanical Pulping Conference, Helsinki, Finland, 6–10 June 1977

Thompson G, Swain J, Kay M, Foster CF (2001) The treatment of pulp and paper mill effluents: a review. Bioresour Technol 77:275–286

Velasco AA, Frostell B, Greene M (1985) Full scale anaerobic-aerobic biological treatment of a semichemical pulping wastewater. Proc. 40th Industrial Waste Conference, Purdue University, West Lafayette, Indiana 1985, pp 297–304

Virkola NE, Honkanen K (1985) Wastewater characteristics. Water Sci Technol 17(1):1–28

Welander T (1988) An anaerobic process for treatment of CTMP effluents. Water Sci Technol 20 (1):143–147

Wiegand et al. (2011) Water profiles of the forest products industry and their utility is sustainability assessment. TAPPI Journal July 2011: 19–27

Chapter 7
Wastewater and Sludge from Pulp and Paper Production Processes

Abstract This chapter presents the quality of the wastewaters from pulping and papermaking operations. Performance data of selected processes and mills are also presented. The characteristics of the wastewater generated from various processes of the pulp and paper industry depend upon the type of process, type of the wood materials, process technology applied, management practices, internal recirculation of the effluent for recovery, and the amount of water to be used in the particular process.

Keywords Wastewater · Wastewater characteristics · Wastewater quality · Sludge · Pulp and paper production · Pulping · Papermaking

The quality of the wastewaters from pulping and papermaking operations are significantly different (Billings and DeHaas 1971) (Table 7.1). This could be due to the diversity of processes and the chemicals used. The major difference between the two is that pulp wastewater contains dissolved wood derived substances which are extracted from the wood during the pulping and bleaching operations (Bajpai 2000). Another difference between the pulp and paper mill effluents is the colour of the final discharge. All pulping effluent has some colour, which is due to the dissolved lignin. This is more pronounced where chemical pulping methods are used. Papermaking effluents also may have some colour, particularly at mills using dyes to produce coloured paper. The pulping process generates a substantial amount of wastewater, approximately 200 m^3/tonne of pulp produced (Cecen et al. 1992), most of which is too weak to recover, although it is highly polluting. In mechanical pulping process, the dissolved organic material from the wood is split between the pulp passing on to the paper machine and that going to waste. The majority of the pollutants which go to the paper machine will be released afterwards in the paper machine wastewater, except in those cases where the process is operated in a closed loop system. In contrast, chemical pulping plants, having recovery, find that most of the organic pollutants dissolved during pulping are retained in the recovered liquors which are usually incinerated. The highest wastewater losses are found in mills using chemi-mechanical process.

© The Author(s) 2017
P. Bajpai, *Anaerobic Technology in Pulp and Paper Industry*, SpringerBriefs in
Applied Sciences and Technology, DOI 10.1007/978-981-10-4130-3_7

Table 7.1 Untreated effluent loads from pulp and paper manufacture

Pulps/paper type	kg/tonne of product	
	Suspended solids	BOD (5 days)
Pulps		
Bleached groundwood	20 ± 38	11 ± 26
Textile fiber	130 ± 220	90 ± 130
Straw	180 ± 220	180 ± 220
De-inked	180 ± 360	26 ± 70
Coarse papers		
Boxboard	22 ± 30	9 ± 18
Corrugating brand	22 ± 30	11 ± 26
Newsprint	9 ± 26	4 ± 9
Insulating board	22 ± 45	67 ± 110
Fine papers	22 ± 45	7 ± 18
Book/publication papers	22 ± 45	9 ± 22
Tissue paper	13 ± 45	4 ± 13

Based on Billings and DeHaas (1971)

Table 7.2 shows the pollutants at various stages of the pulping and paper making process (Pokhrel and Viraraghavan 2004). Individual pulping stage produce different quantities, qualities and types of pollutants. The wastewater pollution load from individual pulping and papermaking process is presented in Table 7.3 (Rintala and Puhakka 1994). The amount of pollutants produced by an individual mill is an important indicator for evaluating the performance of the system and also as a crosscheck whether the mills have followed the guidelines. In Table 7.4, performance data of selected processes and mills are presented (Srivastava et al. 1990; Springer 2000; Vlyssides and Economides 1997).

The wastewater from the papermaking and de-inking process differs from the pulping process because there being no breakdown of raw material, other than the rejects of cleaning and screening (Thompson et al. 2001). The water used in this process plays an important role in controlling the losses of raw material from the wire. This type of wastewater makes up the majority of the effluent released into the water sources in those countries which lack pulping. The wastewater from the de-inking operation contains ink residues which are removed from the waste paper in the flotation de-inking cell and the flotation water clarifier. The sludges do contain heavy metals but these are generally not higher than the levels present in domestic sewage. The papermaking process generates effluent which contains significant quantity of cellulose fines and other additives. This can be up to 50% of the total mass. This contaminated water is usually referred to as whitewater. Reclamation of the effluent is economically important as the gross usage of water in the industry is very high and the cost of effluent treatment for all water assigned to drain would be very expensive, and would also involve a loss of raw materials.

Table 7.2 Pollutants from various sources of pulping and papermaking

Wood Preparation

The soils, dirts, and barks are removed from the wood and chips are separated from the barks and water is used to clean the wood. Thus the wastewater from this source contains suspended solids, BOD, dirt, grit, fibers etc.

Digester house

The waste water generated from the digester house is called "black liquor". Kraft spent cooking "black liquor" contains the cooking chemicals as well as lignin and other extractives from the wood. The wastewater contains resins, fatty acids, color, BOD, COD, AOX, VOCs (terpenes, alcohols, phenols, methanol, acetone, chloroform etc.)

Pulp washing

The wastewater from the pulp washing contains high pH, BOD, COD and suspended solids and dark brown in color

Pulp bleaching

The waste water generated from the bleaching contains dissolved lignin, carbohydrate, color, COD, OX, inorganic chlorine compounds such as chlorate $CLO3-$, Organo chlorine compounds such as dioxins, furans, chlorophenols, VOCs such as acetone, methylene chloride, carbon disulphide, chloroform, chloromethane, trichloroethane etc.

Paper making

The wastewater generated from papermaking contains particulate waste, organic compounds, inorganic dyes, COD, acetone etc.

Based on Pokhrel and Viraraghavan (2004)

Table 7.3 Typical wastewater generation and pollution load from pulp and paper industry

Process	Wastewater (m³/adt pulp or paper)	SS (kg/adt pulp)	COD (kg/adt pulp)
Wet debarking	5–25	nr	5–20
Groundwood pulping	10–15	nr	15–32
TMP-unbleached	10–30	10–40	40–60
TMP-bleached	10–30	10–40	50–120
CTMP-unbleached	10–15	20–50	70–120
CTMP-bleached0	10–15	20–50	100–180
Ca-sulfite (bleached)	150–180	20–60	120–180
Kraft-bleached	60–90	10–40	100–140
Paper making	10–50	nr	nr
Agrobased small paper mill	200–250	50–100	1000–1100
Neutral sulfite semichemicals	20–80	3–10	30–120

Based on Rintala and Puhakka (1994)

Some proportion of the water is recycled to the beaters for use in dilution or other processes.

The characteristics of the wastewater generated from various processes of the pulp and paper industry depend upon the type of process, type of the wood materials, process technology applied, management practices, internal recirculation of the effluent for recovery, and the amount of water to be used in the particular

Table 7.4 Typical pollution load per ton of production (kg/ton)

Process	Pollutants			
	SS	BOD	COD	Color
Wood yard	3.75	1	–	2
Pulping	13.5	5	–	1.5
Bleaching	6	15.5		40
Papermaking	30.8	10.8	–	1.5
Deinking	–	11	54	–
Large mill (India)	31.2	13	82.4	–

Based on Srivastava et al. (1990), Springer (2000), Vlyssides and Economides (1997)

process. The general characteristics of the wastewater produced at various process stages and pollution sources are given in Tables 7.5, 7.6, 7.7 (Welander and Anderson 1985; Jurgensen et al. 1985; Hall and Cornacchio 1988; Velasco et al. 1987; Walters et al. 1988; Frostell 1983; Cocci et al. 1985; Qui et al. 1987; Pipyn 1987).

The anaerobic treatment process requires a certain minimum concentration of degradable organic matter in the effluent if it is to be technically and economically feasible. With the development of new process designs, this limit is gradually decreasing and at present the practical limit is approximately 1000 g/m^3 expressed as COD (Simon and Ullman 1987). This means that for many mills today, particularly older mills with high water consumption, anaerobic treatment may not be quite so appealing. However, in many cases, the concentration can be increased by system closures or separations. It should also be noted that laboratory tests show that the minimum concentration for anaerobic treatment may be reduced further in

Table 7.5 Typical characteristics of wastewater (mg/l) at different processes

Process	Parameters									
	pH	SS	BOD5	COD	Carbohydrate	Acetic acid	Methanol	N	P	S
TMP	–	383	2800	7210	2700	235	25	12	2.3	72
CTMP	–	500	3000–4000	6000–9000	1000	1500	–	–		167
Kraft bleaching	10.1	37–74	128–184	1124–1738	–	0	40–76	–	–	–
Kraft foul	8.0	16	568	1202	–	–	421	–	–	5.9
Sulfite condensate	2.5	–	2000–4000	4000–8000	–	–	250	–	–	800–850
NSSC pulping										
Spent liquor	–	253	13,300	39,800	6210	3200	90	55	10	868
Chip wash	–	6095	12,000	20,600	3210	820	70	86	36	315
Paper mill	–	800	1600	5020	610	54	9	11	0.6	97

Welander and Anderson (1985), Jurgensen et al. (1985), Hall and Cornacchio (1988), Velasco et al. (1987), Walters et al. (1988), Frostell (1983), Cocci et al. (1985), Qui et al. (1987), Pipyn (1987), Bajpai (2000)

Table 7.6 Characteristics of wastewater (mg/l) at various pulp and paper processes

Process	Parameters						
	TS	SS	BOD5	COD	AOX	Resin (µg/l)	Color (Pt–Co)
Wood preparation	1160	600	250	–	–		–
Drum debarking	2017–3171	–	480–987	–	–		20–50
Bleach kraft mill	–	34	23	–	12.5	69	–
Newsprint mill	3750	250	–	3500	–	16	1000

Nemerow and Dasgupta (1991), Springer (2000), Wayland et al. (1998), Tardiff and Hall (1997)

Table 7.7 Characteristics of wastewater (mg/l) at various pulp and paper processes

Process	Parameters					
	pH	TS (mg/l)	SS (mg/l)	BOD5 (mg/l)	COD (mg/l)	Color (Pt–Co)
Digester house	11.6	51,589	23,319	13,088	38,588	16.6a
Combined effluent	7.6	3318	2023	103	675	1.0a
TMP whitewater	4.7	–	91	1090	2440	–
Kraft mill	8.2	8260	3620	–	4112	4667.5
Pulping	10	1810	256	360	–	–
Kraft mill (unbleached)	8.2	1200	150	175	–	25
Bleached pulp mill	7.5	–	1133	1566	2572	4033
Bleaching	2.5	2285	216	140	–	–
Paper making	7.8	1844	760	561	953	Black
Paper mill	8.7	2415	935	425	845	DB
Paper machine	4.5	–	503	170	723	243

Singh et al. (1996), Jahren and Oedegaard (1999), Rohella et al. (2001), Dilek and Gokcay (1994), Yen et al. (1996), Gupta (1997), Dutta (1999)

the future. Certain chemical compounds, mentioned earlier, may disturb the anaerobic process by toxic action and thus make the effluent less suitable for this type of treatment. However, the present trend indicates some possibilities to eliminate interference from these compounds. A large number of studies with pulp and paper industry effluents have been reported (Hall and Cornacchio 1988; Hall 1988; Lee et al. 1989) in the literature during the last few years (Table 7.8). These studies cover many effluent types from debarking, mechanical pulping, deinking, chemimechanical pulping, semichemical pulping, sulphite pulping, sulphate pulping, peroxide bleaching, chlorine bleaching and paper making. Environmental Canada's Wastewater Technology Center screened 42 inplant waste streams from 21 Canadian pulp and paper mills to assess their potential amenability to anaerobic treatment (Hall and Cornacchio 1988). The screening process consisted of chemical characterization and an anaerobic serum bottle technique to demonstrate biodegradability. Twenty three (55%) of the various effluent streams from Kraft, sulphite, mechanical and semichemical mills were found to be suitable for anaerobic treatment (Velasco et al. 1987; Hall et al. 1986; Walters et al. 1988; Frostell 1983; Cocci et al. 1985; Qui et al. 1987; Pipyn 1987). Mechanical and thermomechanical pulping effluents, wastewaters of papermaking and prehydrolysate effluent of dissolving grade pulp mill which are composed predominantly of carbohydrates, are easy to treat anaerobically. Likewise evaporator condensates which are composed mostly of alcohols and volatile fatty acids can be considered as easy for anaerobic treatment. Chemical, semi-chemical and chemithermomechanical pulping liquors, bleaching and debarking effluents are more difficult for anaerobic treatment. These effluents contain important fractions of recalcitrant organic matter and numerous types of toxic compounds. The chemical process generally contribute to the extraction of lignin in the waste water and alkaline chemical conditions lead to solubilization of toxic resin compounds. Bleaching operations often result in the formation of highly toxic chlorinated phenolics. Pulping and bleaching chemicals containing sulphur can contribute to the presence of high concentrations of sulphur in the waste water. The contact of water with bark i.e. wet debarking of wood, produces effluents in which toxic tannic compounds are extracted. The characteristics of some mechanical, chemimechanical, chemical pulping and pulp bleaching effluents that have been successfully anaerobically treated in pilot studies or at full-scale are presented in Table 7.5. A significant fraction of the organics in these effluents are either organic acids or alcohols.

Table 7.8 Effluents used for anaerobic treatment

Kraft
Woodroom
Stripper feed
Contaminated hot water
Evaporator condensate
Sulphite
Neutral sulphite semichemical spent liquor
Final effluent
Clarifier effluent
Combined sewer effluent
Acid condensate (hardwood)
Washer (softwood)
Thermomechanical
Final effluent
Chip wash
Clarifier effluent
Chemithermomechanical pulp
Thermomechanical pulp
Thermomechanical pulp liner board
Nonsulphur semichemical
Controlled effluent
Clarifier effluent

Lee et al. (1989), Hall and Cornacchio (1988)

References

Bajpai P (2000) Treatment of pulp and paper mill effluents with anaerobic technology. Pira International, UK

Billings RM, DeHaas GG (1971) Pollution control in the pulp and paper industry. In: Lund HF (ed) Industrial pollution control

Cecen F, Urban W, Haberl R (1992) Biological and advanced treatment of sulfate pulp bleaching e.uents. Water Sci Technol 26:435 ± 444

Cocci AA, Landine RC, Brown GJ, Tennier AM, Hall ER (1985) Anaerobic treatment of Kraft foul condensates. Proc. Tappi 1985 Environmental Conference, Tappi Press, Atlanta, GA, USA, pp 67–71

Dilek FB, Gokcay CF (1994) Treatment of effluents from hemp-based pulp and paper industry: waste characterization and physicochemical treatability. Water Sci Technol 29(9):161–163

Dutta SK (1999) Study of the physicochemical properties of effluent of the paper mill that affected the paddy plants. J Environ Pollut 6(2 and 3):181–188

Frostell B (1983) Anaerobic pilot scale treatment of a sulphite evaporator condensate. Presented at CPPA 69th Annual Meeting, Montreal, Quebec

Gupta A (1997) Pollution load of paper mill effluent and its impact on biological environment. J Ecotoxicol Environ Monit 7(2):101–112

Hall ER, Cornacchio LA (1988) Anaerobic treatability of Canadian pulp and paper mill wastewaters. Pulp & Paper Canada 89:100–104

Hall ER (1988) Treating wastewater from pulp and paper mills with anaerobic technology. Environmental Technology Notes, published by Environment Canada

Hall ER, Robson RD, Prong CF, Chmelauskas AJ (1986) Evaluation of anaerobic treatment for NSSC wastewater. In: Proceedings of Tappi environmental conference, Atlanta, GA, pp 207–217

Jahren SJ, Oedegaard H (1999) Treatment of thermomechanical pulping (TMP) whitewater in termophilic (55 (C) anaerobic–aerobic moving bed biofilm reactors. Water Sci Technol 40 (8):81–90

Jurgensen SJ, Benjamin MM, Ferguson JF (1985) Treatability of thermomechanical pulping process effluents with anaerobic biological reactors. Proc. Tappi 1985 Environmental Conference, Tappi Press, Atlanta, GA, USA, pp 83–92

Lee JM, Peterson DL, Stickney AR (1989) Anaerobic treatment of pulp mill wastewaters. Env Prog 8(2):73–86

Nemerow NL, Dasgupta A (1991) Industrial and hazardous waste management. Van Nostrand Reinhold, New York

Pipyn P (1987) Anaerobic treatment of Kraft pulp mill condensates. Proc. Tappi 1987 Environmental Conference, Tappi Press, Atlanta, GA, USA, pp 173–177

Pokhrel D, Viraraghavan T (2004) Treatment of pulp and paper mill wastewater—a review. Sci Total Environ 333:37–58

Qui R, Ferguson JF, Benjamin MM (1987) Anaerobic-aerobic treatment of Kraft pulping wastewaters. Proc.Tappi 1987 Environmental Conference, Tappi Press, Atlanta, GA, USA, 1987, pp 165–170

Rintala JA, Puhakka JA (1994) Anaerobic treatment in pulp and paper mill. Bioresour Technol 47:1–18

Rohella RS, Choudhury S, Manthan M, Murthy JS (2001) Removal of colour and turbidity in pulp and paper mill effluents using polyelectrolytes. Indian J Environ Health 43(4):159–163

Singh RS, Marwaha SS, Khanna PK (1996) Characteristics of pulp and paper mill effluents. J Ind Pollut Control 12(2):163–172

Simon O, Ullman P (1987) Present state of anaerobic treatment. Paperi Ja Puu-Paper Och Tra 6:510–515

Springer AM (2000) Industrial environmental control: pulp and paper industry. TAPPI Press, Atlanta, Georgia

Srivastava SK, Bembi R, Singh AK, Sharma A (1990) Physicochemical studies on the characteristics and disposal problems of small and large pulp and paper mill effluents. Indian J Environ Prot 10(6):438–442

Tardif O, Hall ER (1997) Alternatives for treating recirculated newsprint whitewater at high temperatures. Water Sci Technol 35(2–3):57–65

Thompson G, Swain J, Kay M, Forster CF (2001) The treatment of pulp and paper mill effluent: a review. Bioresour Technol 77(3):275–286

Velasco AA, Bonkoski WA, Sarner E (1987) Full scale anaerobic biological treatment of a semichemical pulping wastewater. Proc. Tappi 1987 Environmental Conference, Tappi Press, Atlanta, GA, USA, pp 197–201

Vlyssides AG, Economides DG (1997) Characterization of wastes from a newspaper wash deinking process. Fresenius Environ Bull 6:734–739

Walters JG, Kanow PE, Dalppe HL (1988) A full scale anaerobic contact process treats sulphite evaporator condensate at Hannover. Paper, Alfred, Germany. Proc.Tappi 1988 Environmental Conference, Tappi Press, Atlanta, GA, USA, pp 309–313

Wayland M, Trudeau S, Marchant T, Parker D, Hobson KA (1998) The effect of pulp and paper mill effluent on an insectivorous bird, the tree Swallow. Ecotoxicology 7:237–251

Welander T, Anderson PE (1985) Anaerobic treatment of wastewater from the production of chemithermomechanical pulp. Water Sci Technol 17(1):103–112

Yen NT, Oanh NTK, Reutergard LB, Wise DL, Lan LTT (1996) An integrated waste survey and environmental effects of COGIDO, a bleached pulp and paper mill in Vietnam on the receiving water body. Global Environ Biotechnol 66:349–364

Chapter 8
Anaerobic Treatment of Pulp and Paper Industry Effluents

Abstract The present status of Anaerobic treatment of pulp and paper industry effluents is presented in this chapter. The manufacturers of commercial reactors for waste water treatment and commercial installations are also presented.

Keywords Anaerobic treatment · Pulp and paper industry · Effluents · Manufacturers · Commercial reactors · Waste water treatment · Commercial installations · Forest industry wastewater

8.1 Present Status

Anaerobic technology is being used for the treatment of pulp mill effluents since the middle of 1980s. Earlier, the pulp and paper mill wastewaters were thought too dilute to be treated by the anaerobic process. Development of various high rate anaerobic processes and much more concentrated pulp mill effluents because of the extensive recycling make the economic benefit from anaerobic treatment more significant, which in turn increased the interest in the use of this technology. Anaerobic technologies are already in use for several types of forest industry effluents. Currently several full-scale system are in operation at pulp and paper industries. The most widely applied anaerobic systems are the upflow anaerobic sludge bed (UASB) reactor and the contact process. Most of the existing full-scale anaerobic plants are treating noninhibitory forest industry wastewater which are rich in readily biodegradable organic matter such as recycling waste water, thermomechanical pulping effluents. Full-scale application of anaerobic systems for chemical, semichemical and chemithermomechanical, bleaching and debarking effluents is still limited.

The application of anaerobic treatment for treatment of Kraft bleach plant effluent has been studied by several researchers (Lafond and Ferguson 1991; Raizer-Neto et al. 1991; Rintala and Lepisto 1992). The COD removals have ranged from 28 to 50%. Removal of AOX was improved when easily degradable co-substrate was added to the influent. Several chlorophenolic compounds and chlorinated guaiacols were removed by more than 95% (Parker et al. 1993a, b).

© The Author(s) 2017
P. Bajpai, *Anaerobic Technology in Pulp and Paper Industry*, SpringerBriefs in Applied Sciences and Technology, DOI 10.1007/978-981-10-4130-3_8

Fitzsmons et al. (1990) investigated anaerobic dechlorination/degradation of AOX at different molecular masses in bleach plant effluents. A reduction in AOX was observed with all molecular mass fractions. The rate and extent of dechlorination and degradation of soluble AOX reduced with the increase of molecular weight. As high molecular weight chlorolignins are not amenable to anaerobic microorganisms, dechlorination of high molecular weight compounds may be due to combination of growth, energy metabolism, adsorption and hydrolysis.

Buzzini and Pires (2002) reported 80% average COD removal when treating diluted black liquor from a kraft pulp mill by using an UASB reactor. The performance of a bench scale UASB was also examined by Buzzini et al. (2005) for the treatment of simulated bleached and unbleached cellulose pulp mill wastewaters. They obtained 76% COD removal and 71–99.7% AOX removal. They did not observe any inhibitory effect of the organochlorine compounds on the removal of COD during the experiments.

Chinnaraj and Rao (2006) observed 80–85% reduction in COD, while producing 520 l/kg COD of biogas, after the replacement of an anaerobic lagoon by an UASB installation (full-scale) for the treatment of an agro-based pulp and paper mill wastewater. Furthermore, they obtained a reduction of 6.4 Gg in carbon dioxide emissions through the savings in fossil fuel consumption, and 2.1 Gg reduction in methane emissions from the anaerobic lagoon (equal to 43.8 Gg of carbon dioxide) in nine months.

Zhenhua and Qiaoyuan (2008) obtained 98% reduction in BOD5 and 85.3% reductions in COD from pulping effluents by using a combination of UASB and sequencing batch reactors (SBRs), whereas the removal efficiency when the substrate was just treated by a UASB reactor was considered to be 95% for BOD5 and 75% for COD, at HRT of one day.

Rao and Bapat (2006) observed 70–75 and 85–90% reductions of COD and BOD, respectively, and a methane yield of 0.31–0.33 m^3/kg of COD reduced, when using a full-scale UASB for treating the pre-hydrolysate liquor from a rayon grade pulp mill.

Puyol et al. (2009) used both UASB and anaerobic expanded granular sludge bed reactor (EGSB) for studying the effective removal of 2,4-dichlorophenol. They reported that EGSB reactor showed a better efficiency for the removal of both COD and 2,4-dichlorophenol (75 and 84%, respectively), when compared with UASB reactor (61 and 80%, respectively), at loading rates of 1.9 g COD/l/d and 100 mg 2,4-dichlorophenol/l/d.

Ali and Sreekrishnan (2007) treated black liquor and bleach effluent from an agroresidue-based mill using the anaerobic process. Addition of 1% w/v glucose yielded 80% methane from black liquor with concomitant reduction of COD by 71%, while bleach effluent produced 76% methane and produced 73 and 66% reductions in AOX and COD, respectively. In the absence of glucose, black liquor and bleach effluent produced only 33 and 27% methane reduction with COD reductions of 43 and 31%, respectively.

Thermomechanical pulping of waste water is found to be highly suitable for anaerobic waste water treatment (Sierra-Alvarez et al. 1990, 1991; Jurgensen et al.

1985). In a mesophilic anaerobic process, loading rates up to 12–31 kg COD/m^3/d with about 60–70% COD removal efficiency have been obtained (Sierra-Alvarez et al. 1990, 1991; Rintala and Vuoriranta 1988). In thermophillic anaerobic process conditions, up to 65–75% COD removal was obtained at 55 °C at loading rate of 14–22 kg COD/m^3/d in a UASB reactors (Rintala and Vuoriranta 1988; Rintala and Lepisto 1992).

Kortekaas et al. (1998) studied anaerobic treatment of wastewaters from thermomechanical pulping of hemp. The wood and bark thermomechanical pulping waste waters were treated in a laboratory scale UASB reactor. For both types of wastewaters, maximum COD removal of 72% were obtained at loading rates of 13–16 g COD/l/d providing 59–63% recovery of the influent COD as methane. The reactors provided excellent COD removal efficiencies of 63–66% up to a loading rate of 27 g COD/l/d, which was the highest loading rate tested. Batch toxicity assays showed the absence of methanogenic inhibition by hemp TMP wastewaters, coinciding with the high acetolastic activity of the reactor sludge of approximately 1 g COD/g VSS/d.

The anaerobic treatability of NSSC spent liquor together with other pulping and paper mill waste water streams was studied by Hall et al. (1986) and Wilson et al. (1987). The methanogenic inhibition by NSSC spent liquor was apparently the effect of the tannins present in these wastewaters (Habets and Knelissen 1985). Formation of hydrogen sulfide in the anaerobic treatment of NSSC spent liquor has been reported but this is not related to the methanogenic toxicity. Apparently, the evaporator condensates from the NSSC production are responsive to anaerobic treatment because of their high volatile fatty acid content (Pertulla et al. 1991).

Unstable operations have been observed in anaerobic treatment of pulp mill effluents. The reason for these problems are not clear. It is believed that they may be associated with the toxicants in these effluents (Bajpai 2013).

Research is continuing to develop treatment systems that combine aerobic technology with the ultrafiltration process. The sequential treatment of the effluent from bleached kraft pulp mill in anaerobic fluidised bed and aerobic trickling filters was found to be effective in degrading chlorinated, high and low molecular material (Haggblom and Salkinoja-Salonen 1991). The treatment substantially reduced the COD, BOD and AOX of the waste water. COD and BOD reduction was more in the aerobic process whereas dechlorination was more in the anaerobic process. With the combined aerobic and anaerobic treatment, over 65% reduction of AOX and over 75% reduction of chlorinated phenolics was seen. Measuring the COD/AOX ratio of the wastewater before and after treatment revealed that the chlorinated material was as biodegradable as the non-chlorinated.

Dorica and Elliott (1994) studied the treatability of bleached Kraft effluent using anaerobic plus aerobic processes. BOD reduction in the anaerobic stage was found to vary between 31 and 53% with hardwood effluent. Similarly the AOX removal from the hardwood effluents was higher (65–71%), for the single stage and the two stage treatment respectively, than that for softwood effluents (34–40%). Chlorate was removed easily from both softwood and hardwood effluents (99 and 96% respectively) with little difference in efficiency between the single-stage and two-stage anaerobic systems. At organic loadings rate between 0.4 and 1.0 kg COD/m^3/d, the biogas yields in the reactors were 0.16–0.37 l/g BOD in the feed. Biogas yield was found to be reduced with increasing BOD load for both softwood and hardwood effluents. Anaerobic plus aerobic treatment was able to remove more than 92% of BOD and chlorate. AOX removal was 72–78% with hardwood effluents, and 35–43% with softwood effluents. From hardwood effluents, most of the AOX was found to be removed during feed preparation and storage. Parallel control treatment tests in non-biological reactors confirmed the presence of chemical mechanisms during the treatment of hardwood effluent at 55 °C. The AOX removal that could be attributed to the anaerobic biomass ranged between 0 and 12%. The Enso-Fenox process was found to remove 64–94% of the chlorophenol load, toxicity, mutagenicity and chloroform (Hakulinen 1982).

Haggblom and Salkinoja-Salonen (1991) found the sequential treatment of bleached Kraft effluent in an anaerobic fluidized bed and aerobic trickling filter effective in degrading chlorinated material. The treatment reduced the COD, BOD and the AOX of the waste water. Reduction of COD and BOD was greatest in the aerobic process, whereas dechlorination was significant in the anaerobic process. When the combination of aerobic and anaerobic treatment, was used, over 65% reduction of AOX and 75% reduction of chlorinated phenolic compounds was observed (Table 8.1). Microorganisms capable of mineralizing pentachlorophenol constituted about 3% of the total heterotrophic microbial population in the aerobic trickling filter. Two aerobic polychlorophenol degrading Rhodococcus strains were found to degrade polychlorinated phenols, guaiacols and syringols in the bleaching effluent.

Singh (2007) and Singh and Thakur (2006) investigated sequential anaerobic and aerobic treatment in a two-step bioreactor for removal of colour in the pulp and paper mill effluent. In anaerobic treatment, colour, lignin, COD, AOX and phenol were reduced by 70, 25, 42, 15, 39% respectively in 15 days. The anaerobically treated effluent was separately applied in a bioreactor in presence of a fungal strain,

Table 8.1 Reduction of pollutants in anaerobic-aerobic treatment of bleaching effluent

Parameter	Reduction (%)
COD (mg O$_2$/l)	61
BioCOD (mg O$_2$/)l	78
AOX (mg Cl/l)	68
Chlorophenolic compound	
2,3,4,6 Tetrachlorophenol	71
2,4,6 Trichlorophenol	91
2,4 Dichlorophenol	77
Tetrachloroguaiacols	84
3,4,5 Trichloroguaiacols	78
4,5,6 Trichloroguaiacols	78
4,5 Dichloroguaiacols	76
Trichlorosyringol	64

Based on Haggblom and Salkinoja-Salonen (1991)

Paecilomyces sp., and a bacterial strain, *Microbrevis luteum*. Data indicated reduction in colour, AOX, lignin, COD and phenol by 95, 67, 86, 88, 63% respectively with *Paecilomyces* sp. whereas *M. luteum* showed removal in colour, lignin, COD, AOX and phenol by 76, 69, 75, 82 and 93% respectively by third day when 7 days anaerobically treated effluent was further treated with aerobic microorganisms.

Swedish MoDo Paper's Domsjo Sulfitfabrik is using anaerobic treatment at its sulphite pulp mill and produces all the energy required at the mill (Olofsson 1996). It also fulfills 90% of the heating requirements of the inner town of Ornskoldvik. Two bioreactors at the mill produce biogas and slime from the effluent. The anaerobic unit is used to 70% capacity. Reductions in BOD_7 and COD were 99 and 80% respectively. The slime produced can be used as a fertilizer.

In the Pudumjee Pulp and Paper Mill in India, the anaerobic pretreatment of black liquor reduced COD and BOD by 70 and 90% respectively (Deshpande et al. 1991). The biogas produced is used as a fuel in boilers along with LSHS oil. The anaerobic pretreatment of black liquor has reduced organic loading at the aerobic treatment plant thereby reducing consumption of electrical energy and chemical nutrients.

Swedish researchers reported a process based on ultrafiltration and anaerobic and aerobic biological treatments (EK and Eriksson 1987; EK and Kolar 1989; Eriksson 1990). The ultrafiltration was used to separate the high molecular weight mass, which is relatively resistant to biological degradation. Anaerobic microorganisms more efficiently remove highly chlorinated substances than aerobic microorganisms. The remaining chlorine atoms were removed by aerobic microorganisms. The combined treatments removed 80% of the COD, AOX and chlorinated phenolics and completely removed chlorate (Table 8.2).

In recent years, AnMBRs which combine the advantages of anaerobic digestion process and membrane separation mechanisms are receiving attention because of their advantages for wastewater treatment such as lower energyrequirements and lower sludge production as compared to conventional anaerobic treatment methods (Jeison and Vanlier 2007). Gao et al. (2011) reported that by using anaerobic

Table 8.2 Reduction of pollutants with ultrafiltration plus anaerobic/aerobic system and the aerated lagoon technique

Parameter	UF plus anaerobic/aerobic predicted reductions (%)	Aerated lagoon estimated reductions (%)
BOD	95	40–55
COD	70–85	15–30
AOX	70–85	20–30
Colour	50	0
Toxicity	100	Variable
Chlorinated phenols	>90	0–30
Chlorate	>99	Variable

Based on Eriksson (1990), EK and Eriksson (1987), EK and Kolar (1989)

membrane technologies, it is possible to obtain complete solid–liquid phase separation and, as a result, complete biomass retention. Since 1990s, some studies have been carried out to study the efficiency of such systems for the treatment of pulp and paper mill waste waters, and have shown 50–96% removal of COD (Hall et al. 1995).

Xie et al. (2010) studied the performance of a submerged anaerobic membrane bioreactors (SAnMBRs) for the treatment of kraft evaporator condensateunder mesophilic temperature conditions. They obtained 93–99% COD removal under an organic loading rate of 1–24 kg $COD/m^3/day$. The methane production rate was found to be 0.35 ± 0.05 l/g COD reduced.

Lin et al. (2009) obtained 97–99% COD removal from a kraft evaporator condensate at a feed COD of 10,000 mg/l in two pilot-scale submerged AnMBRs under thermophilic and mesophilic conditions.

Gao et al. (2010) obtained about 90% COD removal during the steady period (22nd–33rd day) of the performance of a submerged AnMBR, treating thermo-mechanical pulping (TMP) whitewater. Several types of membranes such as PVDF based membranes, hollow polymeric fibers, ceramic tubular etc. have been so far developed for the treatment of the various types of wastewaters (Masuelli et al. 2009; Kim et al. 2011; Stamatelatou et al. 2009). However, flat-sheets of polyvinylidine fluoride (PVDF), as a flexible, low weight, inexpensive, and highly nonreactive material, are the major membranes used for the treatment of pulp and paper mill effluents such as Kraft evaporator condensate (Lin et al. 2009) and TMP whitewater (Gao et al. 2010) as internal configurations. The maintenance and operational costs arising from membrane fouling and the frequent cleaning requirement of such hydrophobic polymeric membranes and also being relatively energy intensive are nevertheless considered the main hurdle of such treatment systems dealing with various types of wastewaters. After studying the fouling mechanisms in AnMBRs, Charfi et al. (2012) reported that the cake formation is the main mechanism responsible for membrane fouling in AnMBRs. Such findings were also corroborated by Lin et al. (2009). Although some measures such as feed pre-treatment, optimization of operational conditions, broth properties improvements, and membrane cleaning have already been used for controlling the membrane fouling process (Lin et al. 2013), this issue demands further studies for improving the performance of AnMBR.

Yilmaz et al. (2008) studied the performance of two AFs under mesophilic and thermophilic conditions for the treatment of a paper mill wastewater. No significant differences at OLRs up to 8.4 g COD/l/d was observed. At higher OLRs, slightly better COD removal and biogas production were seen in the thermophilic reactor, which also denotes the effect of the OLR on the performance of the anaerobic digestion process.

Ahn and Forster (2002a) reported that the specific methane production obtained in an anaerobic filter treating a simulated paper mill wastewater under thermophilic temperature was higher than the one obtained at a mesophilic temperature under all the studied HRTs from 11.7 to 26.2 h. They also observed that the performance of the two mesophilic and thermophilic upflow anaerobic filters treating a simulated paper mill wastewater can be affected either by a reduction or an increase in the operating temperature. They showed that the performance of both digesters, in terms of COD removal efficiency and biogas production at an OLR of 1.95 kg $COD/m^3/day$, was negatively affected by a reduction in the operating temperature to 18–24 and to 35 °C for mesophilic and thermophilic digesters, respectively. When the temperature was increased to 55 and 65 °C in mesophilic and thermophilic digesters, respectively, they also observed an immediate reduction in the treatment efficiency (Ahn and Forster 2002b). But, some studies have also shown that anaerobic biomass have a potential for good recovery after undergoing thermal shock (Buzzini and Pires 2002). The effect of the variations in the operating temperature can be affected significantly by the configuration of the reactor. When compared with other high-rate conventional anaerobic digesters AnMBR seems to be more resistant to temperature variation.

Lin et al. (2009) did not observe any significant difference between the thermophilic and mesophilic anaerobic digestion, when treating pulping wastewater by using a pilot-scale SAnMBR. They also observed that the mesophilic SAnMBR can show a better filtration performance in terms of filtration resistance.

Gao et al. (2011) studied the effect of the temperature and temperature shock on the performance of a SAnMBR treating TMP pressate. They found that the COD removal at 37 and 45 °C was slightly higher than that at 55 °C. However, they observed no significant differences between the methane productions at the various temperatures. They also reported that temperature shock can affect the diversity and richness of the species. A COD removal efficiency of 97–99% was observed at a feed COD of 10,000 mg/l in both SAnMBRs. In spite of the advantages of conventional mesophilic and, thermophilic treatments low-temperature anaerobic digestion has emerged in recent years, as an economic method to deal with cool, dilute effluents which were considered as inappropriate substrates for anaerobic digestion (Bialek et al. 2012).

McKeown et al. (2012), by reviewing the basis and the performance of the low temperature anaerobic treatment of wastewater, concluded that the adoption of effective post treatments for low temperature anaerobic digestion is a way to satisfy the stringent environmental regulations. Some recent studies have also indicated that low temperature anaerobic digestion can be more efficient by adopting the co-digestion approach (in pilot-scale application) (Zhang et al. 2013). However, significant physical, chemical and biological improvements should be applied to high-rate anaerobic digestion under low-temperature conditions to enhance the efficiency of the present anaerobic digestion systems, and to improve the amount of the methane produced during the related anaerobic processes.

Anaerobic processes were earlier considered being very sensitive to inhibitory compounds (Lettinga et al. 1991; Rinzema 1988). But now advances in the identification of inhibitory compounds and substances in paper mill effluents and also increasing insight into the biodegradative capacity and toxicity tolerance of anaerobic microorganisms has helped to establish that anaerobic treatment of various inhibitory wastewaters is feasible. The capacity of anaerobic treatment to reduce organic load depends on the presence of significant amounts of persistent organic matter and toxic substances. Most important toxicants are reported below (Pichon et al. 1988; Sierra-Alvarez and Lettinga 1991; McCarthy et al. 1990; Field et al. 1989:

- Sulfate and sulfite
- Wood resin compounds
- Chlorinated phenolics
- Tannins.

These compounds are highly toxic to methanogenic bacteria at a very low concentration. In addition a number of low molecular weight derivatives have also been found as methanogenic inhibitors (Sierra-Alvarez and Lettinga 1991).

In CTMP effluents, volatile terpenes and resins may account for up to 10% of the wastewater COD (1000 mg/l) (Welander and Andersson 1985). The solids present in the CTMP effluent were found to contribute to 80–90% of the acetoclastic inhibition (Richardson et al. 1991). The inhibition caused by resin acids was solved by diluting anaerobic reactor influent with water or aerobically treated CTMP effluent which contained less than 10% of the resin acids present in the untreated wastewater (MacLean et al. 1990; Habets and de Vegt. 1991). Similarly, inhibition by resin acids was solved by diluting the anaerobic reactor influent with water and by aerating the wastewater to oxidise sulfite to sulfate before anaerobic treatment (Eeckhaut et al. 1986).

The AOX generated in the chlorination and alkaline extraction stages are generally considered responsible for a major portion of the methanogenic toxicity in the bleaching effluents (Ferguson et al. 1990; Rintala et al. 1991; Yu and Welander 1994). Anaerobic technologies can be successfully used for reducing the organic load in inhibitory waste waters if dilution of the influent concentration to subtoxic levels is feasible (Lafond and Ferguson 1991; Ferguson and Dalentoft 1991). Dilution prevents methanogenic inhibition and favour microbial adaptation to the inhibitory compounds. Dilution with other non-inhibitory waste streams such as Kraft condensates and sulfite evaporator condensates(Sarner et ai. 1987) before anaerobic treatment, is found to be effective for reducing this toxicity.

Tannic compounds present at very high concentrations are found to inhibit methanogenesis (Field et al. 1988, 1991). Dilution of wastewater or polymerization of toxic tannins to high molecular weight compounds by auto oxidation at high pH as the only treatment (Field et al. 1991) was found to enable anaerobic treatment of debarking effluents.

A system consisting of an anaerobic process followed by an aerobic process appears to be a better option for the removal of COD, AOX and colour from pulp and paper mill effluents (Pokhrel and Viraraghavan 2004). Tezel et al. (2001) reported 91% removal in COD and 58% removal in AOX by using sequential anaerobic and aerobic digestion systems to treat pulp and paper mill wasrewater at a HRT of 5 and 6.54 h for the anaerobic and aerobic processes, respectively.

Bishnoi et al. (2006) obtained a maximum methane production up to 430 ml/day. Furthermore, a COD removal up to 64% was obtained, while volatile fatty acids increased up to 54% at a pH of 7.3, a temperature of 37 °C and 8 days HRT during anaerobic digestion. Afterwards, COD and BOD removals were 81 and 86%, respectively, at 72 h HRT in activated sludge process. It also seems that a combination of fungal and bacterial strains can help for a more effective removal of recalcitrant pollutants from streams. Treatment of the combined effluent of a pulp and paper mill by using a sequential anaerobic and aerobic treatment in two steps bioreactor was studied by Singh and Thakur (2006). They observed 70% reduction in colour, 42% reduction in COD and 39% reduction in AOX in 15 days. However, using a mixture of fungi and bacteria (*Paecilomyces* sp. and *Microbrevis luteum*) for the treatment of anaerobically treated pulp and paper mill effluents, about 95, 67, and 88% reductions in colour, AOX, and COD after 7 and 3 days in the anaerobic and aerobic treatment of the effluents, respectively were observed.

Combination of a UASB reactor (step 1) and two-step sequential aerobic reactor, involving *Paecilomyces* sp. (step 2) and *Pseudomonas syringae* pv *myricae* (CSA105) (step 3), as aerobic inoculums for the treatment of pulp and paper mill effluents, was studied by Chuphal et al. (2005). They found that by using such three-step fixed film sequential bioreactors, 87.7, 76.5, 83.9 and 87.2% removals of colour, lignin, COD, and phenol, respectively, can be obtained.

Balabanic and Klemencic (2011) in a full-scale aerobic and combined aerobic-anaerobic treatment plants, obtained removal efficiencies of 87 and 87% for dimethyl phthalate, 73 and 88% for dibutyl phthalate, 79 and 91% for diethyl phthalate, 84 and 78% for di(2-ethylhexyl) phthalate, 86 and 76% for benzyl butyl phthalate, 74 and 79% for bisphenol A and 71 and 81% for nonylphenol from paper mill effluents, respectively.

Sheldon et al. (2012) conducted a pilot plant study in a EGSB reactor. They reported reduction in COD by 65–85% over a 6 month period. The overall COD removal after the combination of an EGSB with a modified Ludzack–Ettinger process coupled with an ultra-filter membrane was consistent at 96%.

Lin et al. (2014) reported 50–65% COD removal from four different wastewaters from kraft mill using anaerobic process by using a pilot-scale packed bed column at an OLR of 0.2–4.8 kg COD/m^3/d. The overall COD removal after combining with completely mixed activated sludge process, as anaerobic–aerobic sequential system, was found to be 55–70%. The methane production yield was 0.22–0.34 m^3 methane/kg COD, with the biogas containing 80% of methane.

Grover et al. (1999) obtained a maximum of 60% COD removal from black liquor treatment by using an anaerobic baffled reactor at an organic loading rate of 5 kg/m^3/d, a HRT of 2 d, a pH 8.0 and a temperature of 35 °C.

Table 8.3 summarizes the performance of various reactor configurations for the anaerobic treatment of pulp and paper mill wastewaters.

Table 8.3 Anaerobic treatment of pulp and paper mill wastewater

Reactor configuration	Effluents origin	Initial COD (mg/l)	COD removal	References
UASB	Diluted black liquor	1400	76–86	Buzzini et al. (2005)
UASB	Bagasse-based P&P mill	2000–7000	80–85	Chinnaraj and Rao (2006)
UASB + SBR UASB	Wheat straw explosion pulping effluent	–	85.3	Zhenhua and Qiaoyuan (2008)
UASB	Pre-hydrolysate liquor from a rayon grade Pulp mill	2500	70–75 d	Rao and Bapat (2006)
UASB	P&P mill	1133.9 ± 676	∼81	Turkdogan et al. (2013)
SGBRe	P&P mill	1133.9 ± 676	∼82	Turkdogan et al. (2013)
Submerged AnMBR	Kraftevaporator condensate	2500–2700	93–99	Xie et al. (2010)
Submerged AnMBR	TMPwhitewater	2782–3350	90	Gao et al. (2010)
ABR	Recycled paper mill effluents	3380–4930	Up to 71	Zwain et al. (2013)
ABR	Black liquor	10,003 ± 69	60	Grover et al. (1999)

8.2 Manufacturers of Commercial Reactors for Waste Water Treatment and Commercial Installations

Most commercial anaerobic reactors for wastewater treatment are based on the upflow anaerobic sludge blanket (UASB) or internal circulation (IC) reactor principles (Kamali et al. 2016; Zhang et al. 2015). The reactors may also be based on combinations of the special features of different reactors so that their efficiency can be optimized. The commercial manufacturers of anaerobic digesters are listed in Table 8.4. Van lier (2007) has reported that in pulp and paper industry 249 reactors have been installed.

The first full-scale low-rate anaerobic lagoon system for treating paper mill effluents was successfully operated in 1976 by Orient Paper mills, Amlai, India which is an integrated bleached sulphate pulp and paper mill (Dubey et al. 1982) and then in North America in 1978 by the Inland Container Corporation Newport, Indiana (Priest 1980, 1983). In Orient Paper mill, the effluents from washing and screening from the pulp mill and from caustic extraction from the bleach plant are treated. The treatment system is presedimentation-anaerobic lagoon-aerated lagoon-clarification pond (Bajpai 2000). The treatment facility at Inland Container Corporation also has an aerobic polishing step following the anaerobic treatment. The BOD removal was about 85% by anaerobic treatment and 95% by

Table 8.4 Manufacturers of anaerobic digesters	Biothane Systems International, The Netherlands (http://www.biothane.com/en/Biothanetechnologies/Anaerobic-wastewater-treatment)
	Degrémont, France http://www.degremont-industry.com/en/our-expertisetechnologies/wastewater/anaerobic-biological-treatment/
	Paques BV, The Netherlands http://en.paques.nl/pageid=68/BIOPAQ%C2%AE.html
	ADI systems Inc., Canada http://www.adisystemsinc.com/en/technologies/anaerobictreatment
	Purac AB Sweden http://purac.se/?page_id=672
	M/s. Acsion Engineering Pvt. Ltd, India http://www.acsionindia.net/upflow-anaerobic-sludgeblanket.htm
	Clearfleau Ltd. USA http://www.clearfleau.com/page/anaerobic-digestion
	Colsen Group http://www.colsen.nl/csn-prod&serv/en/uasb-ind-enflyer
	Shandong Jinhaosanyang Environmental Protection Equipment. Co., Ltd., China http://www.cnjinhaosanyang.com/cn/product_115_2.html
	Guangxi Bossco Environmental Protection Technology Co., Ltd., China http://www.bossco.cc/newsview-718.aspx
	Based on Zhang et al. (2015)

anaerobic-aerobic treatment. In Hartsville, South Carolina, Sonoco products company's recycle and paper board mill installed a similar anaerobic lagoon and aerobic polish system (Winslow 1988). Gwaliar Rayon mill, Mavoor, India manufactures is treating the prehydrolysis effluent in an anaerobic lagoon (Nambisan et al. 1980). This mill is producing dissolving grade pulp by a prehydrolysis sulphate process. The treatment sequence is neutralization-sedimentation-cooling-anaerobic lagoon treatment and aerated lagoon treatment. Biogas is not collected from the lagoon. About 73% COD removal has been achieved at an influent COD of 80 t/d (flow rate 1700 m^3/d).

The first full-scale application of anaerobic contact systems in the pulp and paper industry was at Swedish sulphite mills in 1983, a semi-chemical pulp and waste paper mill in Spain and a sulphite pulping and cellulose derivative manufacturing facility in Sweden in 1984 and a ground wood mill in Wisconsin in 1986 (Janson 1984; Sarner et al. 1987; Schmutzler et al. 1988). Currently, there are several full-scale anaerobic contact systems in operation at pulp and paper mills worldwide (Bajpai 2000). Reactor volatile solids concentrations reported for anaerobic contact systems operating in the pulp and paper industry have ranged from 3000 to

Table 8.5 Few examples of Using anaerobic technologies in the Pulp and Paper Industry

Mill	Wastewater source	Loading rate (kg COD/m^3/d)	BOD5 (mg/l)	COD (mg/l)
Anaerobic contact reactor				
Hylte Bruk AB, Sweden TMP, groundwood, deink	TMP, groundwood, deinking	2.5	1300	3500
SAICA, Zaragoza, Spain	Waste paper, alkaline cooked straw	4.8	10,000	30,000
Hannover paper, Alfred, Germany	Sulfite effluent condensate	4.2	3000	6000
Niagara of Wisconsin of USA	CTMP	2.7	2500	4800
SCA Ostrand, Ostrand, Sweden	CTMP	6	3700	7900
Alaska Pulp Corporation, Sitka	Sulfite condensate, bleach caustic and pulp white water	3	3500	10,000
Upflow anaerobic sludge blanket				
Celtona, Holland	Pulp whitewater	3	600	1200
Southern paper converter, Australia	Tissue	10	–	10,000
Davidson, United Kingdom	Wastepaper	9	1440	2880
Chimicadel, Friulli, Italy	Linerboard	12.5	12,000	15,600
Quesnel River Pulp, Canada TMP/CTMP	Sulfite	18	3000	7800
Lake Utopia Paper, Canada	Condensate	20	6000	16,000
EnsoGutzeit, Finland Bleached	TMP/CTMP	13.5	1800	4000
McMillan Bloedel, Canada MP	NSSC	15	7000	17,500
Anaerobic filter: Lanaken, Belgium	TMP/CTMP	12.7	4000	7900
Anaerobic fluidized bed: D' Aubigne, France	NSSC/CTMP Paperboard	35	1500	3000

Based on Bajpai (2000)

5000 mg/l to over 10,000 mg/l (Walters et al. 1988; Schmutzler et al. 1988), resulting in volumetric loadings in the range of 1–2 kg BOD removed/m^3/d at BOD removal efficiencies greater than 90% and at optimum temperatures of 35 ± 5 °C. These volumetric loading rates are perhaps 20–50% of those that can be obtained by other high-rate anaerobic treatment configurations.

Since early 1980s, the UASB has been used increasingly in pulp and paper industry (Jain et al. 1998; Habets 1986; Habets and Knelissen 1985; Habets 1986; Rekunen et al. 1985; Habet et al. 1985) and other industries. The loading rates achieved for pulp and paper industry effluents in full-scale UASB plants range from 5 to 27 kg COD/m^3/d. The efficiencies vary from 50 to 80% of the COD depending mostly on the biodegradability of the particular wastewater being treated. The BOD removal efficiencies are high, in most cases between 75 and 99% indicating that anaerobic treatment is particularly useful for the elimination of readily biodegradable organic matter. Several UASB reactors are now operating worldwide for the treatment of pulp and paper mill effluents (Allen and Liu 1998; Rintala and Puhakka 1994). In India, full-scale UASB plants are operating at Harihar Polyfibers and APR Ltd., Satia Paper Mills in Punjab, Warna plant in Maharashtra, India Jain et al. 1998). Table 8.5 presents few examples of using anaerobic technologies in the pulp and paper industry.

References

Ahn JH, Forster C (2002a) The effect of temperature variations on the performance of mesophilic and thermophilic anaerobic filters treating a simulated papermill wastewater. Process Biochem 37:589–594

Ahn JH, Forster C (2002b) A comparison of mesophilic and thermophilic anaerobic upflow filters treating paper–pulp–liquors. Process Biochem 38:256–261

Ali M, Sreekrishnan TR (2007) Anaerobic treatment of agricultural residue based pulp and paper mill effluents for AOX and COD reduction. Process Biochem 36(1–2):25–29

Allen DG, Liu HW (1998) Pulp mill effluent remediation. In: Meyers RA (ed) Encyclopedia of environmental analysis and remediation, vol 6. Wiley, Wiley Interscience Publication, New York, pp 3871–3887

Bajpai P (2000) Anaerobic treatment of pulp and paper industry effluents. Pira Technology Series, UK

Bajpai P (2013) Bleach plant effluents from pulp and paper industry. Springer briefs in applied sciences and technology. Springer International Publishing. doi:10.1007/978-3-319-00545-4

Balabanic D, Klemencic AK (2011) Presence of phthalates, bisphenol A, and nonylphenol in paper mill wastewaters in Slovenia and efficiency of aerobic and combined aerobic-anaerobic biological wastewater treatment plants for their removal. Fresenius Environ Bull 20:86–92

Bialek K, Kumar A, Mahony T, Lens PNL, O'Flaherty V (2012) Microbial community structure and dynamics in anaerobic fluidized-bed and granular sludge-bed reactors: influence of operational temperature and reactor configuration. Microb Biotechnol 5:738–752

Bishnoi NR, Khumukcham RK, Kumar R (2006) Biodegradation of pulp and paper mill effluent using anaerobic followed by aerobic digestion. J Environ Biol 27(2):405–408

Buzzini AP, Pires EC (2002) Cellulose pulp mill effluent treatment in an upflow anaerobic sludge blanket reactor. Process Biochem 38:707–713

Buzzini AP, Gianotti EP, Pires EC (2005) UASB performance for bleached and unbleached kraft pulp synthetic wastewater treatment. Chemosphere 59(1):55–61

Charfi A, Ben Amar N, Harmand J (2012) Analysis of fouling mechanisms in anaerobic membrane bioreactors. Water Res 46(2012):2637–2650

Chinnaraj S, Rao GV (2006) Implementation of an UASB anaerobic digester at bagasse-based pulp and paper industry. Biomass Bioenergy 30(3):273–277

Chuphal Y, Kumar V, Thakur IS (2005) Biodegradation and decolourization of pulp and paper mill effluent by anaerobic and aerobic microorganisms in a sequential bioreactor. World J Microbiol Biotechnol 21(8–9):1439–1445

Deshpande SH, Khanolkar VD, Pudumjee KD (1991) Anaerobic-aerobic treatment of pulp mill effluents—a viable technological option. In: Proceedings of international workshop on small scale chemical recovery, high yield pulping and effluent treatment, Sept 16–20, New Delhi, India, pp 201–213

Dorica J, Elliott A (1994) Contribution of non-biological mechanisms of AOX reduction attained in anaerobic treatment of peroxide bleached TMP mill effluent. In: Proceedings of Tappi international environmental conference, pp 157–163

Dubey RK, Khare A, Kaul SS, Singh MM (1982) Performance of waste treatment plant at Orient Paper Mills, Amlai. International seminar management environmental problems in pulp and paper industry, New Delhi, India, 24–25 Feb 1982

Eeckhaut M, Alaerts G, Pipyn P (1986) Anaerobic treatment of paper mill effluents using polyurethane carrier reactor (PCR) technology. PIRA paper and board division seminar cost effective treatment of papermill effluents using anaerobic technologies, Leatherhead UK, Jan 14–15 1986

EK M, Eriksson KE (1987) External treatment of bleach plant effluent. In: 4th International symposium on wood and pulping chemistry, Paris

EK M, Kolar MC (1989) Reduction of AOX in bleach plant effluents by a combination of ultrafiltration and biological methods. In: Proceedings of 4th International biotech conference in pulp and paper industry, Raleigh, North Carolina, 16–19 May, pp 271–278

Eriksson KE (1990) Biotechnology in the pulp and paper industry. Water Sci Technol 24:79–101

Ferguson JF, Dalentoft E (1991) Investigation of anaerobic removal and degradation of organic chlorine from kraft bleaching wastewaters. Water Sci Technol 24:241–250

Ferguson JF, Luonsi A, Ritter D (1990) Sequential anaerobic/aerobic biological treatment of bleaching wastewaters. In: Proceedings of Tappi 1990 environmental conference. Tappi Press, Atlanta, pp 333–338

Field JA, Leyendeckers MJH, Sierra-Alvarez R, Lettinga G, Habets LHA (1988) The methanogenic toxicity of bark tannins and the anaerobic biodegradability of water soluble bark matter. Water Sci Technol 20(1):219–240

Field JA, Kortekaas S, Lettinga G (1989) The tannin theory of methanogenic toxicity. Biol Wastes 29:241–262

Field JA, Leyendeckers MJH, Sierra-Alvarez R, Lettinga G (1991). Continuous anaerobic treatment of auto-oxidized bark extracts in laboratory-scale columns Biotechnol Bioeng 37:247–255

Fitzsimons R, Ek M, Eriksson K-EL (1990) Anaerobic dechlorination/degradation of chlorinated organic compounds of different molecular masses in bleach plant effluents. Environ Sci Tech 24:1744–1748

Gao WJ, Lin HJJ, Leung KTT, Liao BQQ (2010) Influence of elevated pH shocks on the performance of a submerged anaerobic membrane bioreactor. Process Biochem 45:1279–1287

Gao WJ, Leung KT, Qin WS, Liao BQ (2011) Effects of temperature and temperature shock on the performance and microbial community structure of a submerged anaerobic membrane bioreactor. Bioresour Technol 102:8733–8740

Grover R, Marwaha S, Kennedy J (1999) Studies on the use of an anaerobic baffled reactor for the continuous anaerobic digestion of pulp and paper mill black liquors. Process Biochem 34:653–657

Habets LHA (1986) Experience with the UASB reactor under optimal and suboptimal loadings. PIRA Paper and board division seminar cost effective treatment of paper mill effluents using anaerobic technologies, Leatherhead, England, 14–15 Jan 1986

Habets LHA, de Vegt AL (1991) Anaerobic treatment of bleached TMP and CTMP effluent in the Biopaq UASB system. Water Sci Technol 24:331–345

Habets LHA, Knelissen JH (1985) Anaerobic wastewater treatment plant at Papierfabriek Roermond—working successfully and saving expenses. In: Proceedings Tappi 1985 environmental conference. Tappi Press, Atlanta, pp 93–97

Habets LHA, Tielboard MH, Ferguson AMD, Prong CF, Chmelauskas AJ (1985) Onsite high rate UASB anaerobic demonstration plant treatment of NSSC waste water. Water Sci Technol 20:87–97

Haggblom M, Salkinoja-Salonen M (1991) Biodegradability of chlorinated organic compounds in pulp bleaching effluents. Water Sci Technol (G.B.) 24(3/4):161–170

Hakulinen R (1982) The Enso-Fenox process for the treatment of Kraft pulp bleaching effluent and other waste waters of the forest industry. Paperi Ja Puu-Paper Och Tra 5:341–354

Hall ER, Robson RD, Prong CF, Chmelauskas AJ (1986) Evaluation of anaerobic treatment for NSSC wastewater. In: Proceedings of Tappi environmental conference, Atlanta, GA, pp 207–217

Hall ER, Onysko KA, Parker WJ (1995) Enhancement of bleached kraft organochlorine removal by coupling membrane filtration and anaerobic treatment. Environ Technol 16:115–126

Jain RK, Panwar S, Mathur RM, Kulkarni AG (1998) Biomethanation of pulp and paper mill effluents. Bioenergy News 12:10–12

Janson B (1984) Hylte Braks pioneers in new generation effluent treatment. Pulp Paper Sweden 1984(1):72–76

Jeison D, Vanlier J (2007) Cake formation and consolidation: main factors governing the applicable flux in anaerobic submerged membrane bioreactors (AnSMBR) treating acidified wastewaters. Sep Purif Technol 56(2007):71–78

Jurgensen SJ, Benjamin MM, Ferguson JF (1985) Treatability of thermomechanical pulping process effluents with anaerobic biological reactor. In: Proceedings of Tappi environmental conference. Tappi Press, Atlanta, pp 83–92

Kamali MR, Gameiro T, Costa MEV, Capela I (2016) Anaerobic digestion of pulp and paper mill wastes—An overview of the developments and improvement opportunities. Chem Eng J 298:162–82

Kim MS, Lee DY, Kim DH (2011) Continuous hydrogen production from tofu processing waste using anaerobic mixed microflora under thermophilic conditions. Int J Hydrogen Energy 36:8712–8718

Kortekaas S, Wijngaarde RR, Klomp JW, Lettinga G, Field JA (1998) Anaerobic treatment of hemp thermomechanical pulping wastewater. Water Res 32(11):3362–3370

Lafond RA, Ferguson JF (1991) Anaerobic and aerobic biological treatment processes for removal of chlorinated organics from Kraft bleaching wastes. In: Proceedings of Tappi environmental conference. Tappi press, Atlanta, pp 797–812

Lettinga G, Field JA, Sierra-Alvarez R, vanLier JB, Rintala J (1991) Future perspectives for the anaerobic treatment of forest industry wastewaters. Water Sci Technol 24(3/4):91–102

Lin HJ, Xie K, Mahendran B, Bagley DM, Leung KT, Liss SN, Liao BQ (2009) Sludge properties and their effects on membrane fouling in submerged anaerobic membrane bioreactors (SAnMBRs). Water Res 43:3827–3837

Lin H, Peng W, Zhang M, Chen J, Hong H, Zhang Y (2013) A review on anaerobic membrane bioreactors: applications, membrane fouling and future perspectives. Desalination 314:169–188

Lin C, Zhang P, Pongprueksa P, Liu J, Evers SA, Hart P (2014) Pilot-scale sequential anaerobic–aerobic biological treatment of waste streams from a paper mill. Environ Prog Sustain Energy 33:359–368

Maclean B, de Vegt A, van Driel E (1990) Full scale anaerobic/aerobic treatment of TMP/BCTMP effluent at Quesnel River Pulp Company. In: Proceedings of Tappi 1990 environmental conference. Tappi Press, Atlanta, pp 647–661

Masuelli M, Marchese J, Ochoa NA (2009) SPC/PVDF membranes for emulsified oily wastewater treatment. J Membr Sci 326:688–693

McCarthy PJ, Kennedy KJ, Droste RL (1990) Role of resin acids in the anaerobic toxicity of chemithermomechanical pulp wastewater. Water Res 24:1401–1405

McKeown RM, Hughes D, Collins G, Mahony T, O'Flaherty V (2012) Low temperature anaerobic digestion for wastewater treatment. Curr Opin Biotechnol 23:444–451

Nambisan PNK, Raja KCJ, Mohanchandran TM, Balakrishnan E (1980) Effluent treatment in a rayon grade pulp mill. IPPTA 1980(17):2–10

Olofsson A (1996) Domsjo heats up Ornskoldsvik with biogas. Svensk Papperstidin 99(11):33–34

Parker WJ, Eric R, Farguhar GJ (1993a) Assessment of design and operating parameters for high rate anaerobic fermentation of segregated Kraft mill bleach plant effluents. Water Environ Res 65(3):264–270

Parker WJ, Eric R, Farguhar GJ (1993b) Removal of chlorophenolics and toxicity during high rate anaerobic treatment of Kraft mill bleach plant effluents. Environ Sci Technol 27(9):1783–1789

Pertulla M, Konrusdottin M, Pere J, Kristjansson JK, Viikari L (1991) Removal of acetate from NSSC sulphite pulp mill condensates using thermophilic bacteria. Water Res 25:599–604

Pichon M, Rouger J, Junet E (1988) Anaerobic treatment of sulphur containing effluents. Water Sci Technol 20:133–141

Pokhrel D, Viraraghavan T (2004) Treatment of pulp and paper mill wastewater—A review. Sci Tot Environ 333(1–3):37–58

Priest CJ (1980) A change to anaerobic-aerobic treatment made expanded production possible without expansion of wastewater treatment facilities. In: Proceedings of 35th industrial waste conference, Purdue University, West Lafayette, Indiana, pp 142–146

Priest CJ (1983) Inland container's anaerobic effluent system working well. Pulp Paper 57:125–127

Puyol D, Mohedano AF, Sanz JL, Rodríguez JJ (2009) Comparison of UASB and EGSB performance on the anaerobic biodegradation of 2,4-dichlorophenol. Chemosphere 76:1192–1198

Raizer-Neto E, Pichon M, Beniamin MM (1991) Decreasing chlorinated organics in bleaching effluents in an anaerobic fixed bed reactor. In: Kirk TK, Chang HM (eds) Biotechnology in pulp and paper manufacture (pp 271–278). Butterworth-Heinmann, Stoneham

Rao AG, Bapat AN (2006) Anaerobic treatment of pre-hydrolysate liquor (PHL) from a rayon grade pulp mill: pilot and full-scale experience with UASB reactors. Bioresour Technol 97:2311–2320

Rekunen S, Kallio O, Nystrom T, Oivanen O (1985) The Taman anaerobic process for wastewater from mechanical pulp and paper production. Water Sci Tech 17:133–144

Richardson DA, Andras E, Kennedy KJ (1991) Anaerobic toxicity of fines in chemi-thermome-chanical pulp wastewaters: a batch-assay reactor study comparison. Water Sci. Tech 24 (3/4):103e112

Rintala J, Lepisto S (1992) Anaerobic treatment of thermomechanical pulping wastewater at 35–70 °C. Wat Res 26:1297–1305

Rintala JA, Puhakka JA (1994) Anaerobic treatment in pulp and paper mill. Bioresource Technol 1994(47):1–18

Rintala J, Vuoriranta P (1988) Anaerobic-aerobic treatment of pulping effluents. Tappi J 71:201–207

Rintala JA, Sierra-Alvarez R, Field JA, van Lier JB, Lettinga G (1991) Recent developments in the anaerobic treatment of pulp and paper industry wastewaters. In: Proceedings of Tappi 1991 environmental conference. Tappi Press, Atlanta, pp 777–785

Rinzema A (1988) Anaerobic treatment of wastewater with high concentrations of lipids or sulfate. Doctoral thesis, Dept. Water pollution control, Wageningen, Agricultural University, Wageningen, The Netherlands

Sarner E, Hultman B, Berglund A (1987) Anaerobic treatment using new technology for controlling H_2S toxicity. In: Proceedings of Tappi 1987 environmental conference. Tappi Press, Atlanta, pp 227–232

Schmutzler DW, Eis BJ, Lee JW, Olsen JE (1988) Start up and operation of a full-scale anaerobic treatment system at a groundwood and coated paper mill. In: Proceedings of Tappi 1988 environmental conference. Tappi Press, Atlanta, pp 227–230

Sheldon MS, Zeelie PJ, Edwards W (2012) Treatment of paper mill effluent using an anaerobic/aerobic hybrid side-stream membrane bioreactor. Water Sci Technol 65:1265–1272

Sierra-Alvarez R, Lettinga G (1991) The of wastewater lignins and lignin related compounds. J Chem Technol Biotechnol 50:443–455

Sierra-Alvarez R, Kato M, Lettinga G (1990) The anaerobic biodegradability of paper mill wastewater constituents. Environ Technol Lett 11:891–898

Sierra-Alvarez R, Kortekaas S, vanEckort M, Harbrecht J, Lettinga G (1991) The anaerobic biodegradability and methanogenic toxicity of pulping wastewaters. Wat Sci Tech 24:113–125

Singh P (2007) Sequential anaerobic and aerobic treatment of pulp and paper mill effluent in pilot scale bioreactor. J Environ Biol 28(1):77–82

Singh P, Thakur IS (2006) Colour removal of anaerobically treated pulp and paper mill effluent by microorganisms in two steps bioreactor. Bioresour Technol 97(2):218–223

Stamatelatou K, Kopsahelis A, Blika PS, Paraskeva CA, Lyberatos G (2009) Anaerobic digestion of olive mill wastewater in a periodic anaerobic baffled reactor (PABR) followed by further effluent purification via membrane separation technologies. J Chem Technol Biotechnol 84:909–917

Tezel U, Guven E, Erguder TH, Demirer GN (2001) Sequential (anaerobic/aerobic) biological treatment of Dalaman SEKA pulp and paper industry effluent. Waste Manage 21:717–724

Turkdogan FI, Park J, Evans EA, Ellis TG (2013) Evaluation of pretreatment using UASB and SGBR reactors for pulp and paper plants wastewater treatment. Water Air Soil Pollut 224:1512–1516

Van Lier JB (2007) Current and future trends in anaerobic digestion: diversifying from waste (water) treatment to resource oriented conversion techniques. In: Proceedings of the 11th IWA-international conference on anaerobic digestion, Brisbane, Sept 23–27

Walters JG, Kanow PE, Dalppe HL (1988) A full scale anaerobic contact process treats sulphite evaporator condensate at Hannover (Paper, Alfred, Germany). In: Proceedings of Tappi 1988 environmental conference, Tappi Press, Atlanta, pp 309–313

Welander T, Anderson PE (1985) Anaerobic treatment of wastewater from the production of chemithermomechanical pulp. Water Sci Technol 17(1):103–112

Wilson RW, Murphy KL, Frenelte EG (1987) Aerobic and anaerobic pretreatment of NSSC and CTMP effluent. Pulp Pap Can 88:T4–T8

Winslow FB (1988) Start-up and operation of an anaerobic-aerobic treatment system. Presented at the southern regional meeting, NCASI, New Orleans, June 1988

Xie K, Lin HJ, Mahendran B, Bagley DM, Leung KT, Liss SN, Liao BQ (2010) Performance and fouling characteristics of a submerged anaerobic membrane bioreactor for kraft evaporator condensate treatment. Environ Technol 31:511–521

Yilmaz T, Yuceer A, Basibuyuk M (2008) A comparison of the performance of mesophilic and thermophilic anaerobic filters treating papermill wastewater. Bioresour Technol 99(2008):156–163

Yu P, Welander T (1994) Anaerobic treatment of kraft bleaching plant effluent. Appl Microbiol Biotechnol 40:806–811

Zhang L, Hendrickx TLG, Kampman C, Temmink H, Zeeman G (2013) Codigestion to support low temperature anaerobic pretreatment of municipal sewage in a UASB-digester. Bioresour Technol 148:560–566

Zhang A, Shen J, Ni Y (2015) Anaerobic digestion for use in the pulp and paper Industry and other sectors: an introductory mini-review. BioResources 10(4):8750–8769

Zhenhua S, Qiaoyuan L (2008) Treatment of wheat straw explosion pulping effluent with combined UASB-SBR process. In: 2nd International papermaking and environment, pp 1145–1149

Zwain HM, Roshayu S, Qamaruz N, Abdul H, Dahlan I, Hassan SR, Zaman NQ, Aziz HA, Dahlan I (2013) The start-up performance of modified anaerobic baffled reactor (MABR) for the treatment of recycled paper mill wastewater. J Environ Chem Eng, pp 61–64

Chapter 9
Economic Aspects

Abstract The comparison of the economics of anaerobic and aerobic treatment processes is presented in this chapter. Anaerobic treatment offers substantial savings in electrical power and other benefits which can result in significant savings. These savings are sufficient to offer a return on investment in many cases.

Keywords Economics · Anaerobic treatment · Aerobic treatment · Savings in electrical power · Saving in operating cost · Return on investment

Several researchers have made comparison of the economics of anaerobic and aerobic treatment processes (Habets and Knelissen 1985a, b, c; Eroglu et al. 1994; Maat 1990; Rekunen et al. 1985; Deshpande et al. 1991; Huss et al. 1986).

Habets and Knelissen (1985a, b, c) have reported the economics of many full-scale UASB reactors treating effluents from the paper industry. The pay back of the plant was less than 1.5 years in one board mill. This was due to the significant cost savings from the reduced sewer-discharge levies. In one paper mill at Netherlands-Papierfabriek Roermond, the volumetric capacity of an existing aerobic plant was more than doubled by the addition of BIOPAQ pretreatment plant. Capital and operational costs for complete treatment were reduced by 23% (Habets and Knelissen 1985a). The biogas is used for steam production. Treatment costs per unit of waste treated decreases as loading rates increase.

Anaerobic pretreatment before the existing activated sludge system in the SEKA Bolu Hardboard and Laminated Board (Formica) mill in Turkey provided 70% reduction both in energy for aeration and in excess biological sludge production from the treatment plant (Eroglu et al. 1994). In addition to these, it has been possible to recover bioenergy of 7690 kWh per day, corresponding to 280,000 US dollars per year by adding anaerobic pretreatment into the existing treatment system.

Maat (1990) compared the costs of full-scale anaerobic treatment and aerobic treatment plants. He reported two cases. In one, the capital cost to expand the aerobic system was estimated at 2.5 times the cost for the anaerobic system. For another case, the estimated capital cost for an aerobic system was 2.0 times that for

© The Author(s) 2017
P. Bajpai, *Anaerobic Technology in Pulp and Paper Industry*, SpringerBriefs in Applied Sciences and Technology, DOI 10.1007/978-981-10-4130-3_9

Table 9.1 Cost benefit analysis of aerobic and anaerobic-aerobic treatment[a]

	Conventional aerobic treatment	Anaerobic-aerobic treatment
(a) Capacity to remove BOD kg/d (200 kg BOD/ton bld.pulp)	10,000	10,000
		8500 kg BOD in AN
		1500 kg BOD in AE
(b) Power requirement kwh/d (1.2 kwh/kg BOD removed)	12,000	3500 1 kwh/5 kg BOD in AN
		1.2 kwh/5 kg BOD in AE
(c) Nutrient chemicals		
– Urea (46% N) kg/d	900	450
– DAP (20 N:20 P) kg/d	500	300
(100 BOD:5 N:1 P)		200:5:1 in AN
Removal ratio		100:5:1 in AE
(d) Treated effluent quality		
– BOD_5 20 °C mg/l	80–100	60–70
– COD mg/l	1200–1500	800–1000
(e) BOD_5 removal efficiency %	90–92	97–98
(f) COD removal efficiency %	75–78	83–85
(g) Biogas generation m^3/d (with 70% methane content)	Nil	12,000
		AN-anaerobic
		AE-aerobic
Estimated running expenses and benefits		
(a) Power Rs./day (at Rs. 1.75/kwh)	21,000	6125
(b) Nutrient chemicals Rs./d (at Rs. 5.0/kg)	7,000	3750
(c) Biogas energy kcal/d (at 6000 kcal/m^3)	Nil	72 million
(d) Oil (LSHS equivalent) (1000 m^3 = 0.59 MT)	Nil	7 MT
(e) Value of oil Rs/d (at Rs. 5200/MT)	Nil	36,400
(f) Net benefit Rs./d	28,000	26,525
	(Negative)	(Positive)

Annual Benefits Rs. 54.525 × 330 days = Rs. 180 lacs

[a]For a 50 tpd bleached pulp mill using soda process and /straw as raw material
Based on data from Deshpande et al. (1991)

an equivalent anaerobic system. The space requirement for anaerobic treatment in each case was less than 50% of the requirement for aerobic treatment. A net difference of $0.18 kg BOD removed between anaerobic and aerobic plant operating costs was observed at these two plants. This means an actual cost difference for a typical plant with 30 tonnes of BOD removal capacity of $5400/d or an annual savings of about $1,800,000.

Table 9.2 Savings in operating cost per year at the Hylte Bruk and SAICA ANAMET plants compared to aerobic treatment of the same wastewater

Cost item	Hylte Bruk (US$)	SAICA (US$)
Electrically US$0.02 kwh^{-1}	20,000	230,000
Nutrients	70,000	500,000
Sludge handling	14,000	100,000
Biogas production	130,000	1,170,000
Total savings per year	234,000	2,000,000

Based on Huss et al. (1986)

Rekunen et al. (1985) compared the economic advantages of the TAMAN process with those of the activated sludge process. There was not much difference in the investment, but the operating costs of the TAMAN process was lesser than 10% of those for the same treatment capacity with the activated sludge process. The difference is because of the aeration energy required by the activated sludge plant and the greater nutrient and the polymer requirement. The fuel value of the biogas is nearly FIM 1 million, which effects a similar reduction in TAMAN operating costs. The total costs of the activated sludge plant was more than 2 times the costs of TAMAN.

Table 9.1 shows the cost benefit analysis of full-scale anaerobic-aerobic treatment plant at Pudumjee Pulp and Paper mills in India (Deshpande et al. 1991). It gives typical analysis and comparison for a 50 tpd bagasse/straw based bleached pulp mill. The annual benefit from anaerobic-aerobic combined treatment against only aerobic amount to Rs. 180 lacs. Even, if one estimates a capital cost of Rs. 500 lacs for anaerobic-aerobic plant, the pay-back is very attractive.

Table 9.2 presents the operating cost savings per year obtained in Hylte Bruk and SAICA ANAMET anaerobic-aerobic treatment plants compared to aerobic treatment of the same amounts of BOD (Huss et al. 1986). At the Hylte Bruk mill (investment US$1:5 million), anaerobic treatment saves operating costs but savings are comparatively marginal to the investment whereas at the SAICA mill (investment US$4 million), the savings are significant.

References

Deshpande SH, Khanolkar VD, Pudumjee KD (1991) Anaerobic-aerobic treatment of pulp mill effluents—a viable technological option. In: Proceedings of international workshop on small scale chemical recovery, high yield pulping and effluent treatment, 16–20 Sept. New Delhi, India, 1991:201–213

Eroglu V, Oztruk I, Ubay G, Demir I, Korkurt, EN (1994) Feasibility of anaerobic pretreatment for the effluents from hardboard and laminated board industry. Water Sci Technol 29(5–6): 391–397

Habets LHA, Knelissen JH (1985a) Anaerobic pretreatment reduces mill costs, improves effluent quality. Pulp Paper 59:66–69

Habets LHA, Knelissen JH (1985b) Application of the UASB reactor for anaerobic treatment of paper and board mill effluent. Water Sci Technol 17:61–75

Habets LHA, Knelissen JH (1985c) Anaerobic wastewater treatment plant at Papierfabriek Roermond—working successfully and saving expenses. In: Proceedings of Tappi 1985 environmental conference. Tappi Press, Atlanta, pp 93–97

Huss L, Sievert P, Särner E (1986) Anamet full scale anaerobic treatment experiences at three pulp and paper mills. Pira paper and board division seminar cost effective treatment of paper mill effluents using anaerobic technologies, Leatherhead UK, Jan 14–15

Maat DZ (1990) Anaerobic treatment of pulp and paper effluents. In: Proceedings of Tappi 1990 environmental conference. Tappi Press, Atlanta, pp 757–759

Rekunen S, Kallio O, Nystrom T, Oivanen O (1985) The Taman anaerobic process for wastewater from mechanical pulp and paper production. Water Sci Tech 1985(17):133–144

Chapter 10
Conclusion and Future Perspectives

Abstract Anaerobic treatment systems are viable technologies for wastewater pollution control in the pulp and paper industry and can be used as an essential part of an integrated treatment resource preservation system. The anaerobic treatment system include certain pretreatments (precipitation and oxidative pretreatments) that eliminate or detoxify aromatic compounds or modify aromatic compounds to improve their anaerobic biodegradability.

Keywords Anaerobic treatment · Wastewater pollution control · Pulp and paper industry · Resource preservation system · Pretreatment · Precipitation pretreatment · Oxidative pretreatment · Detoxification · Anaerobic biodegradability

Anaerobic treatment systems are viable technologies for wastewater pollution control in the pulp and paper industry and can be used as an essential part of an integrated treatment resource preservation system (Zhang et al. 2015; Hubbe et al. 2016). Advantages of anaerobic treatment are net production of renewable energy (biogas), reduced biosolids production and reduced emission of greenhouse gases. As a means of waste treatment, and apart from the energy content and value of the biogas, anaerobic treatment offers substantial savings in electrical power and other benefits which can result in significant savings. In many cases, it has been found that these savings are sufficient to offer a return on investment. For pulp and paper industry wastewaters, sulphate and sulphite reduction during anaerobic digestion can be combined with sulphur recovery systems. Particularly, biological recovery of elemental sulphur from hydrogen sulphide in anaerobically treated pulp and paper industry effluents has a promising future (Bajpai et al. 1999). The anaerobic method under thermophilic conditions offers attractive potentials for hot pulp and paper industry effluents allowing the application of higher organic loading rates and eliminating the need for cooling.

Anaerobic treatment has also the potential to remove environmentally harmful organochlorine compounds which are generated during chlorine bleaching of chemical pulps. The removal of AOX by two stage anaerobic-aerobic treatment is

P. Bajpai, *Anaerobic Technology in Pulp and Paper Industry*, SpringerBriefs in Applied Sciences and Technology, DOI 10.1007/978-981-10-4130-3_10

much higher than a single stage anaerobic and aerobic treatment alone (Hubbe et al. 2016; Bajpai 2000, 2013).

Various anaerobic reactors/digesters are commercially available, which include anaerobic lagoon/covered lagoon reactor, stirred reactor/contact reactor, plug-flow anaerobic reactor, anaerobic filter reactor, upflow anaerobic sludge blanket reactor, expanded granular sludge bed reactor, and internal circulation reactor, among others. These reactors can be specifically tailored for practical applications dealing with various feedstocks. A number of producers of these anaerobic reactors are available on the global market. Much potential do exist in terms of the more efficient and widespread use of anaerobic digestion technologies, which calls for technological advancements and breakthroughs related to biochemical, biological, and processing machinery aspects of the process. Future anaerobic digestion technologies such as those related to the concept of integrated biorefinery would play a significant role in meeting the high demand of environmental protection and bioenergy production. The enhancement of the efficiency of anaerobic reactors through scientific and technological innovations would also serve as the key to more widespread commercial use of anaerobic digestion.

In terms of the noticeable advantages of EGSB and IC, the systems show higher resistance to impact, higher organic loading, up-flow velocity and sufficient attachment between sludge and biomass (Mao et al. 2015). With respect to the sustainability of biogas technology, developing optimal cost-optimal input/output ratio of digestion process could be a promising technology. Anaerobic digesters having internal settlers such as UASB reactors are the main reactor systems for the treatment of pulp and paper mill wastewaters. These reactors have shown a moderate to high performance to reduce the COD and various removal efficiencies for other parameters including TSS, BOD, AOX, etc., depending on the operating conditions, reactor design, and the properties of the streams (Kamali et al. 2016).

The results obtained in the full-scale plants show that anaerobic technology should be definitely recognized as a viable alternative to aerobic treatment in a number of cases. In almost all pulp and paper industry, full-scale applications, anaerobic treatment is followed by aerobic post-treatment (Hubbe et al. 2016; Pokhrel and Viraraghavan 2004). The performance of anaerobic-aerobic treatment is superior or at least identical to aerobic treatment (Kamali et al. 2016). The suitability and cost of the anaerobic-aerobic and aerobic treatment systems are largely affected by a variety of mill-specific factors.

In order to facilitate the anaerobic treatment of difficult pulp and paper industry waste waters, numerous measures can be taken involving either the operation or design of the wastewater treatment system. These measures can be used to surpass the previously discussed limitations confronting the application of anaerobic treatment technologies to pulp and paper industry effluents.

During the operation of the bioreactors, anaerobic bacteria should be acclimatized to toxic organic compounds. The wastewaters should be diluted to subtoxic concentrations during reactor start-up and the dilution of the wastewater should be reduced in an incremental fashion in accordance with the degree by which the microorganisms adapt to the toxicity and develop capacities to degrade the organic compounds. Additionally, immobilizing anaerobic bacteria and maintaining high concentrations of biomass in the reactor, are factors which are known to improve the tolerance to toxic substances by anaerobic treatment systems.

The design of the anaerobic treatment system can include certain pretreatments that eliminate or detoxify aromatic compounds or otherwise modify aromatic compounds to improve their anaerobic biodegradability. These pretreatments include precipitation pretreatments and oxidative pretreatments. Precipitation pretreatments have been applied to anaerobic wastewater treatment systems to either remove toxic compounds or remove recalcitrant fractions.

At higher temperatures, biochemical reactions can proceed more rapidly which implicates a decrease of retention times or a smaller reactor volume of the wastewater treatment plants. Many effluents of pulp and paper industry are discharged at high temperatures which make them attractive for thermophilic treatment as no further energy input is required. Thermophilic treatment is shown to be suitable for several types of wastewaters. High-strength wastewaters can be treated at very high loading rates with high COD removal efficiency. At high temperature, the liquid viscosity is lower, which might benefit the biomass hold-up in upflow reactors if low strength wastewater is treated. To ensure the highest process stability, excess temperature fluctuations should be prevented. The activity of thermophilic sludge decreases with decreasing temperature but is still considerable and comparable with mesophilic sludge at temperatures of 40–45 °C. This is due to the fact that the growth rate and maintenance consumption of thermophilic bacteria is 2–3 times higher than the mesophilic counterparts. Startup of the thermophilic reactors was demonstrated to be easy and can be done by increasing the reactor temperature directly to the desired level or following a gradual increase with only a few degrees difference during several months. Besides temperature, process stability is also dependant on the chosen reactor type. Systems with a high biomass retention tend to be less sensitive than completely mixed reactors because of the higher variety of available selection criteria and thus variety of biomass, inside the reactor.

References

Bajpai P (2000) Anaerobic treatment of pulp and paper industry effluents. Pira Technology Series, UK

Bajpai P (2013) Bleach plant effluents from pulp and paper industry. In: SpringerBriefs in Applied sciences and Technology. Springer International Publishing. doi:10.1007/978-3-319-00545-4

Bajpai P, Bajpai PK, Kondo R (1999) Biotechnology for environmental protection in pulp and paper industry. Springer, Germany

Hubbe MA, Metts JR, Hermosilla D, Blanco MA, Yerushalmi L, Haghighat F, Lindholm-Lehto P, Khodaparast Z, Kamali M, Elliott A (2016) Wastewater treatment and reclamation: a review of pulp and paper industry practices and opportunities. BioResources 11(3):7953–8091

Kamali MR, Gameiro T, Costa MEV, Capela I (2016) Anaerobic digestion of pulp and paper mill wastes—an overview of the developments and improvement opportunities. Chem Eng J 298:162–182

Mao C, Feng Y, Wang X, Ren G (2015) Review on research achievements of biogas from anaerobic digestion. Renew Sustain Energy Rev 45:540–555

Pokhrel D, Viraraghavan T (2004) Treatment of pulp and paper mill wastewater—a review. Sci Total Environ 333(1–3):37–58

Zhang A, Shen J, Ni Y (2015) Anaerobic digestion for use in the pulp and paper industry and other sectors: an introductory mini-review. BioResources 10(4):8750–8769

Index

A

Acidification, 9, 45, 46

B

Bagasse, 88, 89
Bagasse-based P&P mill, 78

C

Chlorophenolic compound, 69, 72
Chlorophenols, 63
Chronic toxicity, 58
Clarifier, 66
Clarifier effluent, 66
Cleaning, 55–57, 62, 74
Clostridium, 9
Coarse papers, 62
Cobalt, 21
COD, 2, 18, 21, 22, 30, 32, 33, 39, 41–45, 58,
    63–65, 70–77, 79–81, 88, 92, 93
Colloidal material, 49
Colour, 30, 58, 61, 72, 73, 76, 77
Combined sewer effluent, 66
Contaminated hot water, 66
Contaminated water, 62
Controlled effluent, 66
Conventional aerobic
    treatment, 88
Conventional digesters, 38
Cooking, 56, 63
Cooling, 55, 79, 91
Corrosive, 10, 18
Corrugated paper, 57
Corrugating brand, 62
Corynebacterium, 9
CSTR system, 37
CTMP, 64, 76, 80
CTMP-bleached, 63
CTMP-unbleached, 63

D

DAP, 88
Debarking, 31, 56, 65, 66
Debarking effluent, 69, 76
Dechlorination, 70–72
Dehydrogenation, 7
De-inked, 57, 62
Deinking, 64, 65, 80
Desulfobacter, 9
Desulfomonas, 9
Desulfovibrio, 9
Detoxification, 19
Dibutyl phthalate, 77
2,4 dichlorophenol, 72
Di(2-ethylhexyl) phthalate, 77
Diethyl phthalate, 77
Digestate, 7
Digester, 11, 14, 16, 17, 20, 22–24, 38, 41, 43,
    57, 75, 78, 79, 92
Digester house, 63, 65
Dimethyl phthalate, 77
Dioxins, 57, 63
Dirt, 63
Dissolved lignin, 58, 61, 63
Dissolved wood, 61
Dissolving grade pulp mill, 66
Distilleries, 3
Downflow reactor, 44
DTPA, 18, 21

E

Effluent, 16, 17, 19, 20, 24, 31, 40, 46, 57, 61,
    62, 65, 70–72, 76, 77, 79, 88
Electrical power, 33, 91
Electron acceptor, 10, 33
Elementary Chlorine Free (ECF), 57
Enanthate, 19
Enanthic acid, 19

© The Author(s) 2017
P. Bajpai, *Anaerobic Technology in Pulp and Paper Industry*, SpringerBriefs in
Applied Sciences and Technology, DOI 10.1007/978-981-10-4130-3

Energy, 2, 3, 10, 15, 20, 25, 30–33, 40, 43, 73,
    74, 87, 89, 91, 93
Energy metabolism, 70
Environmental impact, 58
Enzymatic hydrolysis, 9
Escherichia coli, 9
Evaporator condensate, 57, 66, 71, 74, 76, 78
Expanded Granular Sludge Bed Reactor
    (EGSB), 37, 46, 70

F
Facultative, 9, 19, 38, 50
Fats, 7
Fatty acids, 2, 8, 11, 15, 17, 19, 49, 66, 77
Fermentation, 7
Fermentative microorganisms, 8
Fertilizer, 29, 73
Filtration resistance, 75
Final effluent, 66
Fine papers, 62
Fish, 31, 58
Fixed biofilm, 50
Flocculant anaerobic biomass, 50
Flocculant seed, 50
Flocs, 42
Flotation de-inking, 62
Flotation water clarifier, 62
Fluidized bed reactor, 37
Food processing, 3, 38
Footprint, 2
Forest biorefinery, 2
Formic acid, 9
Fossil fuel, 1, 70
Fouling mechanism, 74
Furans, 57, 63

G
Gallotannic acid, 21
Gas agitation system, 49
Glucose, 70
Glycerol, 8
Granular sludge, 3, 37, 38, 42, 43, 50, 70, 92
Granules, 41, 46
Grease, 39
Grindstone, 56
Grit, 63
Groundwood pulping, 63
Guaiacol, 69, 72

H
Hardboard, 87
Hardwood, 66, 71, 72
Hardwood effluent, 71
Headbox, 57

Heat power, 1
Heat recovery, 56
Heavy metals, 20, 62
Hemp, 71
High strength wastewaters, 3
Hollow polymeric fibers, 74
Hybrid reactor configurations, 50
Hydraulic Retention Time (HRT), 24, 39, 43
Hydrogen, 1, 9–11, 15, 17–21, 50
Hydrogenophilic, 10
Hydrogen-scavenging bacteria, 9
Hydrogen sulfide, 71
Hydrolysis, 7–9, 11, 15, 70
Hydrolytic exo-enzymes, 8
Hydrophobic, 74

I
Inhibitory effect, 70
Inorganic dyes, 63
Insulating board, 62
Integrated biorefinery, 2, 92
Integrated pulp and paper mill, 57
Internal circulation reactor, 37, 46
Iron, 19–21

K
Kraft, 10, 58, 63, 65, 66, 70–72, 74,
    76–78
Kraft-bleached, 63
Kraft bleaching, 64
Kraft bleach plant effluent, 69
Kraft evaporator condensate, 74
Kraft foul, 64
Kraft mill, 65
Kraft spent cooking, 63

L
Lactobacillus, 9
Large mill, 64
Lipase, 8
Lipolytic bacteria, 11
Liquefaction, 7
Long chain fatty acids, 8
LSHS oil, 73

M
Macronutrient, 21
Magnesium, 21
Mechanical pulp, 19, 21
Mechanical pulping, 56, 58, 61, 65
Membrane, 18, 37, 39, 49, 73, 74, 77
Membrane cleaning, 74
Membrane coupled high-rate, 37
Membrane fouling, 49, 74

Membrane surface, 49
Mesophilic, 2, 14, 15, 32, 71, 74, 75, 93
Mesophilic methanogens, 23
Mesophilic temperature, 14, 74
Methane, 1, 3, 7, 9–11, 14–16, 18, 20, 21, 23, 24, 29–33, 40, 50, 70, 74, 75, 77
Methane bacteria, 10
Methane-producing bacteria, 10
Methanobacterium formicicum, 9
Methanobrevibacter ruminantium, 9
Methanococcus vannielli, 9
Methanogenesis, 10, 15, 21, 76
Methanogenic bacteria, 9, 18, 19, 76
Methanogenic toxicity, 21, 31, 71, 76
Methanogenium cariaci, 9
Methanogens, 9–11, 15, 16, 18–24, 31, 32
Methanol, 9, 10, 18, 63, 64
Methanomicrobium, 9
Methanosarcina barkei, 9
Methanospirilum hungatei, 9
Methylene chloride, 63
Methyltrophic bacteria, 10
Micrococcus, 9
Micronutrients, 21
Microorganisms, 7, 9, 11, 16, 21–24, 29, 31, 32, 39, 44, 58, 70, 72, 73, 75, 92
Mineral filler, 57
Mixing, 13, 24, 39, 43
Molecular hydrogen, 20
Molecular oxygen, 13, 57
Molybdenum, 21
Municipal wastewater, 49
Mutagenicity, 72

N
Neutralization, 79
Neutral Sulphite Semichemical (NSSC), 10
Neutral sulphite semichemical spent liquor, 66
Newsprint, 57
Nickel, 20, 21
Nonsulphur semichemical, 66
Nonylphenol, 77

O
Obligate, 9, 10, 13, 19
Oil, 39, 88
Open earthen basin, 38
Optical brighteners, 57
Organic acids, 9, 16, 17, 66
Organic fouling, 49
Organic loading rate, 18, 30, 32, 39, 41, 43, 74, 77, 91
Organic polymer, 7
Orthophosphate, 22

Oxygen, 1, 2, 7, 9, 10, 18, 19, 29, 38, 40, 57, 58
Oxygen demand, 10
Ozone, 57

P
Packing material, 44
Paper machine white water, 57
Paper making, 2, 56, 57, 62, 63, 65
Paper mill effluent, 1, 29, 61, 72, 74–78, 81
Pathogen, 11, 15, 23, 24, 43
Peptococcus, 9
Peroxide, 17, 19, 57, 65
Peroxide bleaching, 21
Phenols, 21, 30–32, 63
Phosphate salts, 22
Phosphoric acid, 22
Pollution control, 91
Pollution load, 9, 57, 58, 62–64
Polychlorinated phenols, 72
Polymeric membrane, 74
Polymerization, 76
Polyphenolic polymer, 30
Polyvinyl alcohol, 57
Polyvinylidine fluoride, 74
Porosity, 44
Prehydrolysate effluent, 66
Prehydrolysate liquor, 70
Pressure, 10, 15, 17, 20, 56, 57
Pretreatment, 19, 73, 87
Pretreatment systems, 38
Propionate, 10, 18–20
Propionic acid, 9, 19
Propionic acid degrading bacteria, 15
Protease, 8
Proteins, 7
Proteolytic, 11
Pseudomonas, 9, 77
Psychrophilic, 14
Pulp and paper industry, 2, 3, 18, 30, 38, 55, 57, 58, 63, 65, 78–81, 91–93
Pulp bleaching, 56, 57, 63, 66
Pulping, 19, 30, 56–58, 61–63, 65, 66, 69, 71, 75
Pulping operation, 56, 57, 61
Pulps, 62, 91
Pulp washing, 57, 63

R
Raw material preparation, 56, 57
Rayon grade pulp mill, 70
Recalcitrant organic matter, 66
Recycled fibre, 57
Refining, 56

Renewable energy, 1, 91
Resin acids, 21, 30, 76
Resin compounds, 21, 30, 31, 66, 76
Resins, 63, 76
Retention time, 10, 24, 93

S
Saccharolytic, 11
Sarcina, 9
Screening, 56, 57, 62, 78
Screening process, 65
Secondary clarifier, 39
Sedimentation, 44, 79
Selemonas, 9
Selenium, 21
Semichemical pulping, 65
Sequencing batch reactors, 70
Sequential anaerobic and aerobic treatment, 72, 77
Settling tank, 39, 40
Sizing agent, 10
Slime, 73
Sludge, 1–3, 10, 19, 29, 37–45, 47, 62, 69, 71, 73, 77, 87, 89, 92, 93
Soda process, 58, 88
Sodium, 21
Softwood, 66, 71
Softwood effluents, 72
Soil conditioning, 29
Solid–liquid phase separation, 74
Solids retention time, 22, 43
Solubilisation, 7
Specific surface area, 44, 45
Specific water consumption, 57
Spent pulping liquor, 57
Staphylococcus, 9
Starch, 7, 57
Stock preparation, 57
Straw, 62, 78, 80, 88, 89
Streptococcus, 9
Strict anaerobes, 9, 10, 19, 50
Stripper feed, 66
Submerged anaerobic membrane bioreactors, 74
Sugars, 8, 17, 19
Sulfate, 17, 31, 40, 76
Sulfite, 17, 76
Sulfite condensate, 64, 80
Sulfite evaporator condensates, 76
Sulfur, 21
Sulphate, 10, 17, 18, 78, 79, 91
Sulphate pulping, 65
Sulphate-reducing bacteria, 10, 18, 33
Sulphide, 10, 17–20, 91

Sulphite, 10, 17–19, 65, 66, 91
Sulphite pulping, 65, 79
Sulphite pulp mill, 73, 79
Sulphur-reducing bacteria, 10
Suspended solids, 39–42, 46, 62, 63
Sustainable energy, 1
Synthrophy, 10
Syntrophic relationship, 10, 20
Syringol, 72

T
Tannic compounds, 31, 66, 76
Tannin, 21, 30–32, 71, 76
TCF bleached pulp, 57
Temperature, 11, 13–15, 23, 43, 57, 74, 75, 77, 80, 93
Terpenes, 21, 30, 63, 76
Tetrachloroguaiacols, 72
2,3,4,6 tetrachlorophenol, 72
Textile fiber, 62
Thermomechanical, 66
Thermomechanical pulping, 64–66, 69–71, 74, 78, 80
Thermomechanical pulping waste water, 70, 71
Thermomechanical pulp liner board, 66
Thermophilic, 14, 18, 32, 74
Thermophilic digester, 14, 75
Thermophilic process, 15
Thermophilic reactor, 74, 93
Thermophilic treatment, 75, 93
Thickening, 56, 57
Thiosulphate, 18
Tissue paper, 62
Titanium dioxide, 57
TMP-bleached, 63
TMP pressate, 75
TMP-unbleached, 63
Totally Chlorine Free (TCF), 57
Toxic, 3, 9, 10, 16–22
Toxicity, 13, 18, 21, 31, 32, 50, 72, 73, 75, 76, 92
Transportation fuel, 1
Trichloroethane, 30, 63
3,4,5 trichloroguaiacols, 72
4,5,6 trichloroguaiacols, 72
2,4,6 trichlorophenol, 72
Tungsten, 21

U
Ultrafiltration, 71, 73
Upflow anaerobic sludge blanket reactor, 41
Upflow reactor, 93
Urea, 22, 88

V
Valerate, 19
Valeric acid, 19
Veillonella, 9
VOCS, 63
Volatile solids, 23, 79
Volatile solids loading rate, 21, 23, 24
Volatile terpenes, 21, 30, 76

W
Washer, 57, 66
Washing, 57, 63, 78
Waste stream, 1, 2, 65, 76
Waste water, 2, 29, 57, 58, 63, 66, 69, 71, 72,
    74, 76

Wastewater treatment, 3, 10, 15, 33, 42, 73, 78,
    92, 93
Water-insoluble organics, 9
Wet debarking, 63, 66
Wheat straw explosion pulping effluent, 85
White grades, 57
White water, 57, 80
Wood, 18, 21, 30, 44, 56, 61–63, 65, 66, 71, 79
Wood debarking, 56
Wood preparation, 63
Wood resin compounds, 76
Wood room, 66
Wood yard, 64

Printed in the United States
By Bookmasters